COMMERCIAL APPLIANCES

REFERENCE MANUAL

G25

Published by ConstructionSkills, Bircham Newton, King's Lynn, Norfolk, PE31 6RH

© **Construction Industry Training Board 2001**

The Construction Industry Training Board otherwise known as CITB-ConstructionSkills and ConstructionSkills is a registered charity (Charity Number: 264289)

First published 2001
Revised February 2003
Revised December 2005
Revised August 2007

ISBN: 978-1-85751-253-3

ConstructionSkills has made every effort to ensure that the information contained within this publication is accurate. Its content should be used as guidance material and not as a replacement for current regulations or existing standards.

All rights reserved. No part of this publication may be reproduced, stored in a retrieval system or transmitted in any form or by any means, electronic, mechanical, photocopying, recording or otherwise, without the prior permission in writing from ConstructionSkills.

Printed in the UK

For a comprehensive listing of all BES publications turn to the back page.
Tel: 01485 577800   Fax: 01485 577758   E-mail: publications@cskills.org

CONTENTS

	Page
WARM AIR HEATING	1
Indirect Fired Air Heaters	1
Direct Fired Air Heaters	4
Installation Requirements	5
Heater Location	5
Gas Pipework	6
Ventilation	6
Flues	7
RADIANT HEATING	9
Systems	10
High Intensity	10
Low Intensity	10
Radiant Plaque Heaters	10
Lever, Cock and Chain Type	11
Automatic	11
Radiant Cone Heaters	11
Portable Floor Standing Heaters	12
Catalytic Radiant Panels	13
Radiant Tube Heaters	13
Continuous Radiant Tube Systems	14
Air Heated Radiant Tube Systems	14
Installation Requirements	15
Heater Location	16
Gas Pipework	17
Ventilation	18
Flues	18
WATER BOILERS AND SYSTEMS	19
Natural Draught (Conventional 'Open' Flue)	20
Natural Draught (Balanced Flue)	21
Fan Draught (Forced or Induced – Closed Flue)	22
Fan Upstream of the Burner – Forced Draught	22
Fan Downstream of the Burner – Induced Draught	23
Fan Dilution Systems	24
Modular Boiler Systems	25
High Efficiency Modular Boiler Systems	27
Condensing Boilers	28
Storage Water Heaters	29

(continued overleaf)

	Page
Cast Iron Sectional Boilers	30
Steel Shell Boilers	31
Installation Requirements	32
Location	32
Gas Pipework	33
Ventilation	34
Flues	34
Balanced Compartment or High Level Supply Discharge	35
Ancillary Equipment	36
COMMISSIONING GAS EQUIPMENT	37
Inspection Period	41
Dry Run	42
Live Run	43
Commissioning	44
SERVICING	48
Pre-Service Checks	49
Full Service	50
GENERAL FAULT FINDING	52
Fault Finding Checklist for Commercial Appliances with Natural Draught Burners and Thermo-Electric F.F.D.	52
Fault Finding Checklist for Commercial Appliances with Forced/Induced Draught Burners and Automatic Control	56
Fault Finding Checklist for Commercial Air Heaters	60
Fault Finding Checklist for Commercial Boilers	70

WARM AIR HEATING

Warm air heating systems for the industrial and commercial market sector fall into two categories; indirect fired and direct fired.

Both systems utilise the principle of forced convection warm air circulation to heat a given work space or warehouse area. There is a wide variation of individual suspended or floor standing units dependent upon heat output requirements.

Indirect fired air heaters separate the combustion process from the heated air supply by means of a combustion chamber and heat exchanger, with exhaust gases passing to external atmosphere through a connected flue pipe. The combustion process heats these to a preset temperature before air to be heated is blown across by a fan unit forcing the warmed air into the room space.

A variation of this type of system is the warm air curtain whereby the fanned outlet air is ducted over or at the side of a doorway to blast the air across the open door. This will provide a warm air barrier to heat incoming cold air and minimise heat loss through the open door.

Direct fired air heaters take combustion air direct from outside atmosphere, with or without make up air from within the room, and forces the fanned air directly over a flame and into the room to be heated. Products of combustion are diluted by fresh air directly drawn in before being discharged into the room space, i.e. the heater unit is unflued.

INDIRECT FIRED AIR HEATERS

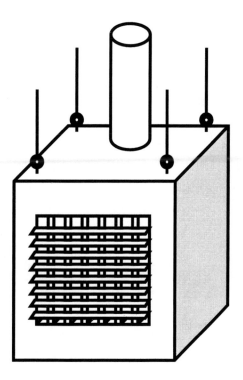

Suspended warm air unit

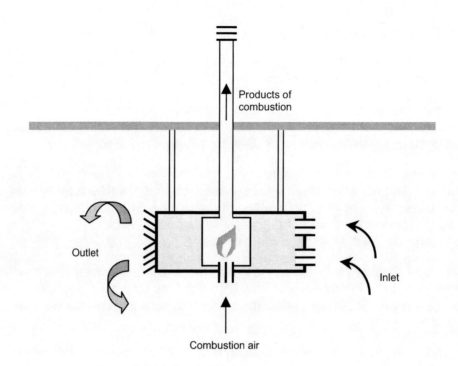

Indirect unit air heater

Indirect fired heaters are manufactured in a range of units with various heat outputs from approximately 10 kW to 440 kW. They can be floor standing or suspended from the building structure, although output capacity of standard suspended units will generally be limited to a nominal maximum of 140 kW.

Traditionally, the suspended unit will be under-fired using a natural draught burner system with products of combustion rising through the heat exchanger to a flue exit at the top of the heater. An axial or centrifugal fan will be mounted at the rear of the heater, which blows air across the heat exchanger through adjustable outlet louvres or grilles to discharge into the room.

Suspended units offer significant advantages when factory or retail floor space is at a premium as they will be located unobtrusively at high level. However, a disadvantage could be in the increased number of units required compared with a fewer number of the larger floor standing units.

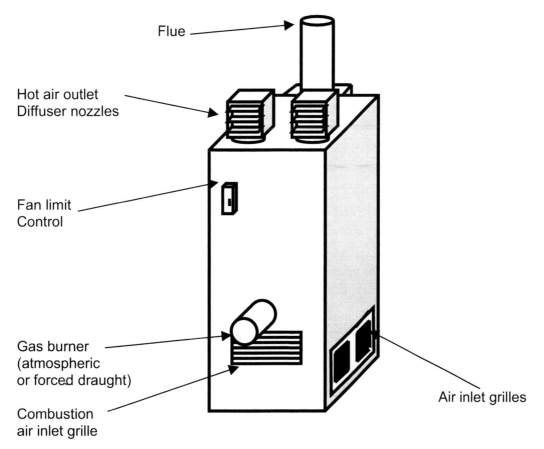

Floor standing warm air unit

Floor standing units are usually free standing due to their weight and size. Traditionally units up to a nominal 100 kW capacity may utilise natural or forced draught burners, above 100 kW will generally utilise forced draught burners only, which fire into the combustion chamber. Products of combustion then pass through a separate heat exchanger before being passed to the flue system. A 230/400 V centrifugal fan unit is located under the combustion chamber and heat exchanger to blow air across to the air outlet nozzles. Multiple nozzles can usually be moved to provide 360° spread of warm air to all parts of the room. Alternatively, ducted models are available for multiple room heating applications and improved heat distribution.

DIRECT FIRED AIR HEATERS

Direct fired gas heating is very efficient. Cold air is drawn from outside the building, heated to the required temperature and blown directly into the space to be heated.

There is no heat exchanger, no flue, and no wasted fuel. The total thermal efficiency of the whole system is high – over 90%.

Because fresh air is being heated the environment is kept fresh so there is no need for additional ventilation or the heat losses associated with indirect fired systems.

With a direct-fired system, the design is based on an intent to slightly pressurise the building so that draughts blow outwards, not in. Also temperature stratification is reduced because the air is designed to distribute gently and evenly.

This type of system is particularly suitable for applications where there is a contaminated atmosphere from an industrial or commercial process such as catering, swimming pools, paint spray booths and welding workshops, etc., where a high rate of change with fresh air is required.

In the aforementioned applications the contaminated air in the room may be discharged through an extract system. Any such exhaust system should be interlocked with the heater to prevent its operation in the event of airflow failure.

Any direct fired heating system must be correctly designed to ensure that the products of combustion are not allowed to build up in the occupied room space.

Main threshold values are: 10 ppm Carbon monoxide (CO)
 2800 ppm Carbon dioxide (CO_2)

CO_2 is present in the atmosphere at a typical background level of 300 ppm, any heater system will need to be designed to limit any increase to a maximum of 2500 ppm. This will be achieved if the system works on 100% fresh air intake and the temperature rise limited to 55°C.

Some systems are fitted with CO_2 limiting controls. These systems must inherently fail-safe and must be regularly calibrated.

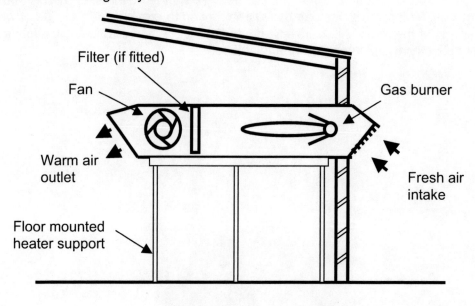

Direct fired warm air unit

INSTALLATION REQUIREMENTS

Warm air heaters must be installed in accordance with the relevant provisions of the Gas Safety (Installation and Use) Regulations and obligations that arise as a result of the Health and Safety at Work Act.

When installing gas fired heating systems, guidance should be sought from appropriate reference standards. The specific relevant reference standards are as follows:

BS 6230: 2005 Specification for Installation of Gas – forced convection air heaters for industrial and commercial space heating (2nd family gases). *(N.B. 2nd family gases = natural gas)*

IGE/UP/10 Edition 3 Installation of Gas Appliances in Industrial and Commercial Premises

MANUFACTURER'S INFORMATION

Manufacturer's recommendations should always be consulted prior to installation work commencing in order to establish all particular requirements, but these are likely to cover the following criteria.

Heater Location

Any appliance needs to be located in a safe and secure manner where it will not be damaged by other activities, for example the movement of fork lift trucks or overhead cranes.

Typically, unit heaters may be suspended from roof trusses but may also be fixed on purpose made brackets from a wall. Any means of suspension or support must be of sufficient number and strength to support the full weight of the heater and the unsupported part of its flue system.

Floor standing heaters need to be stable or rigidly fixed.

Any ducting from the heater needs to be as direct and short as possible with due regard to satisfactory distribution of the heated air, providing a level of warmth without adversely affecting personnel, for example, by excessive fanned air currents.

The floor on which any heater may be sited must not be subjected to temperatures in excess of 65°C when the heater is in operation. Adequate clearance must be provided between any combustible materials, or any material likely to be affected by high temperatures, and the appliance or flue. (Manufacturer's information will usually provide guidance on minimum clearances.)

As with any gas appliance in the commercial and industrial environment, consideration must be given to the location of the heaters and the atmosphere where they are to be used. Heaters must not be installed within buildings that have been classified as hazardous areas. The definitions for hazardous areas can be found in BS EN 60079-10.

Heaters supplying warm air to hazardous areas need to be located either in a separate room/compartment or in the open air. If the heater is sited in a compartment adjacent to an area where flammable heavier than air vapours may be present, the compartment itself will be subject to certain requirements, for example a 450 mm high non-combustible door sill, self closing door and one hour fire resistance, etc. Other conditions may be applicable which affect the heater position, such as location of outlet ducts 1.8 m above floor level and fresh air intake requirements.

Siting the heater in garage workshops may present problems due to the presence of petroleum and other heavier than air vapours. In this environment the heater controls (time switches, thermostats) must not be sited within 1.2 m of the floor level within the space to be heated. The heaters themselves must be sited at least 1.8 m above the floor level.

Any heater must be suitable for the conditions in which it is to be installed, particularly if sited externally. Alternative materials for finishes and manufacture may be necessary in corrosive environments. This also applies to the installation materials such as suspension chains, gas pipe and flexes.

Gas Pipework

The general pipework installation should comply with the 'Institution of Gas Engineers' reference standard UP/2 – Gas Installation pipework, boosters and compressors on industrial and commercial premises. A guide to commercial pipework and testing is available from ConstructionSkills (G24).

The gas supply pipework needs to be adequately sized to provide the working pressure for the volume required by the heater(s) to maintain the nominal heat input.

The Gas Safety (Installation & Use) Regulations requires that a suitable isolation valve be fitted to the appliance. A pipe union will usually be required downstream of the appliance isolation valve to permit burner removal for maintenance purposes.

Plant pipework upstream of the burner needs to be adequately supported, making allowance for any vibration and/or movement. There may be a requirement for suspended heaters to be connected to the gas supply system using a corrugated metallic gas flex manufactured to BS 6501-1, manufacturer's advice should be sought.

Where such a flex is required it needs to be installed in a manner that does not result in any strain or torsion and a 90° action manual valve will be required upstream of the flexible pipe assembly.

Ventilation

Adequate ventilation must be provided to permit the safe operation of the appliance(s).

BS 6230 and the commercial ventilation section of the ConstructionSkills reference manual (G23 or G88) gives detailed information on calculating air supply requirements for any given heating application.

Where heaters are located within a factory or similar type building applications, normal building volume air change rates are usually sufficient to provide adequate air for combustion. Additional ventilation grilles therefore need only be fitted where there is uncertainty as to the actual volume of air change, i.e. if it is likely to be less than half an air change per hour.

Where natural ventilation is provided, grilles need to be located at low level (below the appliance flue connection).

Ventilation openings need to be sited where they cannot easily be blocked or flooded, ideally on at least two sides of the building and located so that suction of the exhaust air is not disturbed by wind influence.

When considering exhaust air openings the airflow rate needs to accommodate requirements for other purposes. The size and number of openings will be based on the higher of these airflow rates.

Flues

Requirements for flue systems are to be found in BS 6230, IGE/UP/10, manufacturer's instructions, and the commercial flues section of the ConstructionSkills reference manual (G23 or G88). For appliances exceeding 150 kW the 3rd edition of the 1956 Clean Air Act Memorandum also applies regarding flue termination and minimum flue gas velocity.

Any flue system needs to be constructed of non-combustible materials that are not unduly affected by heat, condensation and the products of combustion.

Manufacturer's information should be consulted to ensure that maximum lengths of flue pipe are not exceeded. Where exhaust fans are fitted the resistance of the flue system must not be greater than the capability of the exhaust fan.

Also of particular importance for some types of heater is the need to ensure that the flue can be easily disconnected to facilitate correct maintenance procedures and repairs.

An approved type terminal will be required on flue pipe systems up to 200 mm diameter, but above this may not be necessary.

To overcome some conventional flue installation difficulties e.g. if there is an additional floor level immediately above the installation and a vertical flue would be impossible, some manufacturer's provide additional flue facility options such as room sealed (illustrated) to enable short horizontal flue systems for suspended air heaters.

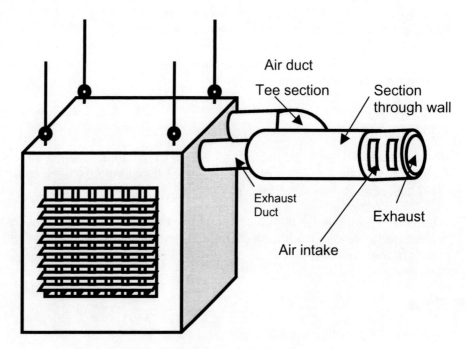

Room sealed suspended warm air unit

Or, for longer horizontal flue systems, some mechanical assistance such as an exhaust fan will be required. It is essential that any such mechanical exhaust system be linked to the operation of the gas burner such that should the exhaust fan system fail the burner cannot operate.

RADIANT HEATING

Radiant heating systems are chosen because they may offer certain advantages over other types of system, such as centralised boiler wet systems or warm air systems.

Rather than heating the total air volume of the building, radiant heaters project infra-red radiation to a chosen area. Heaters may also reflect heated air off solid bodies in the direction of the chosen area.

Radiant energy can therefore be targeted to the people in the work area and also warm adjacent solid bodies, creating a climate whereby occupants feel warm although the actual air temperature may be cool. This effect would be similar to that of sunshine on a cool day in spring or autumn.

Areas popularly heated by radiant means include warehouses, factories, garage workshops, showrooms, churches, sports halls, DIY stores and garden centres.

There are two main types of radiant systems:

- high intensity (luminous) – radiant ceramic plaque and cone heaters
- low intensity (black heat) – radiant tubes.

The type of system adopted will depend on the particular requirements of the area to be heated. High intensity heaters are flueless and so will be better suited for spot heating applications in large, well-ventilated areas. Radiant tubes can be used as part or complete heating systems and may be flued or unflued.

Advantages of radiant heating systems are:

- heat is efficiently directed to a chosen area by one heater or evenly distributed by multiple heaters and an arrangement of 'cross fields'. This allows selected areas to be heated, e.g. in a factory where only certain areas may have people at work

- infra-red radiation heats surfaces not air volumes. The air is only heated when reflected or convected from a heated surface

- radiant heating will minimise the effect of heat loss through the building fabric, especially upper parts of a typical building construction, e.g. factories, commercial units and warehouses

- reduced warm up time. Radiant heaters rapidly reach their normal operating temperature so personnel operating beneath them quickly feel the benefit of the radiated heat, although the surrounding air temperature and building structure may still be cool

- they will often present energy savings compared to other systems

- radiant heaters are generally quiet in operation and can be located in most cases exactly where they will be most effective and efficient

- heaters are generally suspended or wall mounted and therefore do not occupy valuable factory floor space.

SYSTEMS

High Intensity

- Plaque:
    - lever cock and chain, piezo ignition, thermo-electric
    - automatic.
- Cone.
- Floor Mounted (Portable).

Low Intensity

- Catalytic Panels.
- Radiant Tubes
    - linear
    - 'U' tube
    - continuous
    - air heated (direct fired duct system).

RADIANT PLAQUE HEATERS

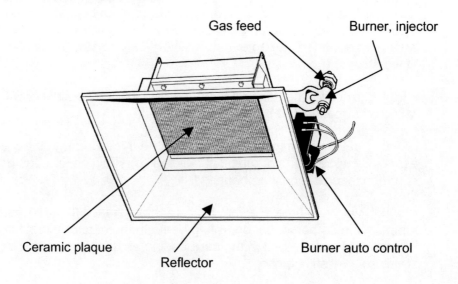

Radiant plaque heater

The luminous ceramic plaque is usually a combination of a number of standard sized ceramic panels built up and contained in a metal frame to provide a variety of heat inputs ranging from 3 to 30 kW approximately.

An air/gas mixture is supplied from an atmospheric injector to a series of perforations formed in the ceramic panel. The surface temperature of the panel when it is at its normal operating condition is approximately 850°C and is the reason they are denoted 'high intensity'.

The mounting height of the heaters will directly affect the radiant intensity. The lower the mounting height from floor level the greater the radiant intensity will be per square metre.

Manufacturer's Information will specify minimum mounting height, typically 3.7 m but larger heater units need to be mounted higher to ensure human comfort. Radiation intensity 240 W/m² should be considered a maximum at head height. Higher levels of radiant intensity will cause people to feel unwell from the 'hot head/cold feet' syndrome.

Lever, Cock and Chain Type

Traditional heaters were turned 'on' and 'off' by a simple level valve with chains brought down to low level.

Automatic

Newer appliances are more likely to be automatic or semi-automatic and may work in conjunction with an energy management system.

RADIANT CONE HEATERS

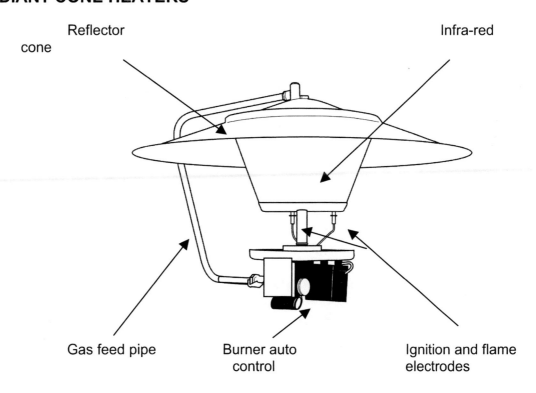

Radiant cone heater

This heater utilises a ring burner which fires onto a ceramic fibre gauze, surrounded by a perforated stainless steel emitter. A circular aluminium reflector deflects the radiated heat downwards.

They are particularly suited for small workshops, individual work areas and incorporated in patio heaters. They are available in various heat outputs between 5 and 22 kW approximately.

Usually available with permanent pilot thermo-electric or fully automatic flame ignition and safeguard systems.

PORTABLE FLOOR STANDING HEATERS

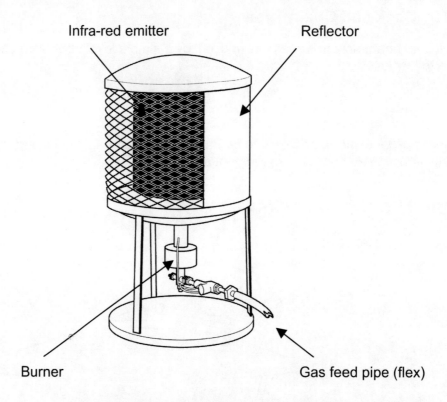

Portable floor standing infra-red heater

Similar in principle to the cone heater but designed for portable heating applications.

The units will therefore usually utilise LPG as the fuel source due to the need for portability, but are available for natural gas firing. They are available in various heat outputs between 16 and 50 kW approximately.

CATALYTIC RADIANT PANELS

Natural gas catalytic panels are not widely used in the UK but can be found on certain types of process plant such as paint or leather drying. Such processes may use solvents or gases with low ignition point and catalytic panels could be used without risk of causing explosions. However, they are more commonly found in conjunction with LPG for domestic and camping space heaters.

The general principle relies on gas passing through a porous pad impregnated with a catalyst. Oxygen from the atmosphere will then diffuse through the pad in the reverse direction causing a chemical reaction at the catalyst surface and generate heat at approximately 450°C without the appearance of a flame.

There has been a major problem in the past associated with 'methane slippage' on natural gas catalytic panels.

RADIANT TUBE HEATERS

Non-luminous or 'black heat' radiant tube heaters work by the internal heating of a mild or heavy duty mild steel tube which in turn emits radiated heat directed by a polished aluminium reflector shield. A heavy-duty tube can vary from 6 m to 11 m in length and 75 mm to 100 mm in diameter for appliances of input up to 40 kW.

Temperature emitted from the tube will vary from 400 to 500°C at the burner end, reducing to approximately 200°C at the exhaust end.

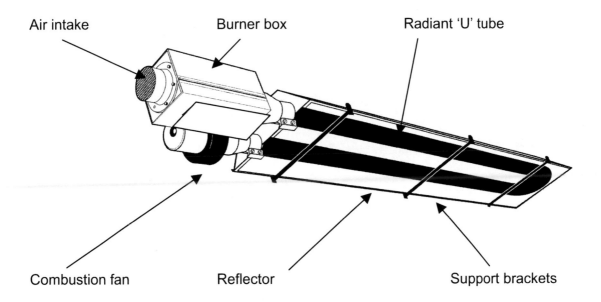

Radiant tube heater

Single burner tube heaters may be linear or 'U' tube configuration with induced or forced draught and may incorporate a flue break or draught diverter.

Groups of individual tube heaters can be served by manifold flueing to an external wall, which is usually served by a vacuum pump.

Continuous Radiant Tube Systems

The continuous radiant tube system will be designed to provide total heating to a particular area. It comprises a number of linear tube burners connected in series along a common tube. Burners are located at between 3.5 m to 6 m intervals dependant on individual burner ratings and the required radiant intensity level.

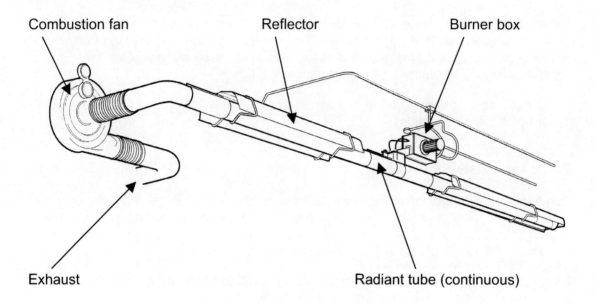

Continuous radiant tube heater

The tube system terminates with a suction fan, which draws the products of combustion through the entire tube system and exhausts to outside atmosphere.

A multi legged system will need to be balanced correctly by dampers fitted at the end of each leg to ensure the correct air pressure at each individual burner.

Gas throughput is controlled by a governor or a zero governor (dependant upon the manufacturer) at each burner so that constant gas/air ratio is maintained.

The entire tube heating system is fitted with automatic control, incorporating air purge, air-flow proving device, spark ignition and flame monitoring.

Air Heated Radiant Tube Systems

The air heated radiant tube system consists of a direct-fired air heater connected to a large diameter continuous closed loop duct.

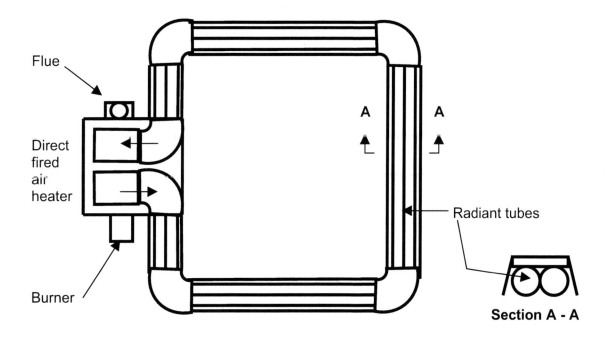

Air heated large bore radiant tube system

The ducts may be 600 mm diameter in banks of up to three looped circuits.

INSTALLATION REQUIREMENTS

Radiant heaters must be installed in accordance with the relevant provisions of the Gas Safety (Installation and Use) Regulations and obligations that arise as a result of the Health and Safety at Work Act.

When installing radiant heating systems guidance should be sought from appropriate reference standards. The <u>specific</u> relevant reference standards are as follows:

BS EN 13410: 2001	Gas fired overhead heaters – ventilation requirements for non-domestic premises
BS 6896: 2005	Specification for installation of gas fired overhead radiant heaters for industrial and commercial heating
BS 7186: 1989	Specification for non-domestic gas fired overhead radiant tube heaters
IGE/UP/10 Edition 3	Installation of gas appliances in industrial and commercial premises

MANUFACTURER'S INFORMATION

Manufacturer's recommendations should always be consulted prior to installation work commencing in order to establish all particular requirements, but these are likely to cover the following criteria.

Heater Location

Any appliance needs to be located in a safe and secure manner where it will not be damaged by other activities, for example, the movement of fork lift trucks or overhead cranes.

Typically, plaque and tube heaters may be suspended from roof trusses but may also be fixed and angled on a wall. Any means of suspension or support must be of sufficient number and strength to support the full weight of the heater and the unsupported part of its flue system.

Any heater needs to be located at a suitable height where satisfactory levels of warmth will be emitted without resulting in excessive radiation levels. Thermal radiated intensity levels of 80 W/m² to the floor level and 240 W/m² at head heights must not be exceeded.

Combustible materials in the vicinity of any heater or its flue must not be subjected to temperatures in excess of 65°C when the heater is in operation. Adequate clearance must therefore be provided between any combustible materials, or any material likely to be affected by high temperatures, and the appliance or flue. (Manufacturer's information will usually provide guidance on minimum clearances.)

It is important that reflectors and diverter shields be positioned correctly, not only to direct the radiation but also to position outlets correctly for dispersal of the products of combustion from plaque heaters.

Heater location should be made with consideration to avoiding shadowing by persons or structures and the most effective coverage over the heating area.

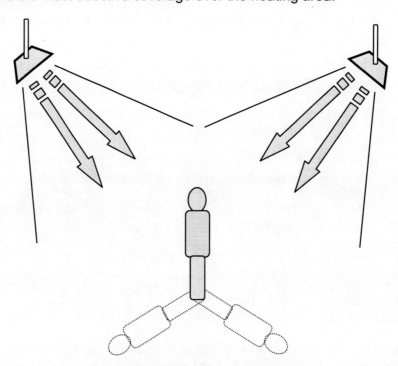

Heaters arranged to avoid shadowing

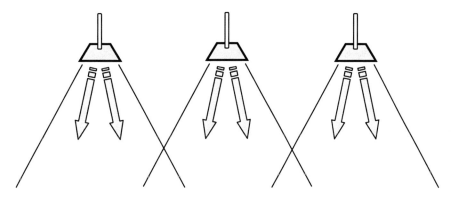

Heaters arranged for even heat distribution

As with any gas appliance in the commercial and industrial environment, consideration must be given to the location of the heaters and the atmosphere where they are to be used.

Radiant tube heaters may be fitted with a ducted fresh air intake for combustion air if the heater is to be sited in a dust-laden atmosphere.

Any heater must be suitable for the conditions in which it is to be installed, particularly in corrosive and high wind speed environments. Alternative materials for finishes and manufacture may be necessary in corrosive environments. This also applies to the installation materials such as suspension chains, gas pipe and flexes.

Siting the heater in garage workshops may present problems due to the presence of petroleum and other heavier than air vapours. In this environment the heater controls (time switches, thermostats) must not be sited within 1.2 m of the floor level within the space to be heated and the heater itself must be in excess of 1.8 m from the floor. In any event hazardous areas must include a risk assessment in accordance with BS EN 60079-10: 2003 before installation.

Gas Pipework

The general pipework installation should comply with the 'Institution of Gas Engineers' reference standard UP/2 – Gas Installation Pipework, Boosters and Compressors on Industrial and Commercial Premises.

The gas supply pipework needs to be adequately sized to provide the working pressure for the volume required by the heater(s) to maintain the nominal heat input.

The Gas Safety (Installation and Use) Regulations requires that a suitable isolation valve be fitted to the appliance. A pipe union will usually be required downstream of the appliance isolation valve to permit burner removal for maintenance purposes.

Plant pipework upstream of the burner needs to be adequately supported, making allowance for the vibration and movement associated with radiant tube heaters particularly. It is normally a requirement for suspended tube heaters to be connected to the gas supply system using a corrugated metallic gas flex manufactured to BS 6501-1 or 669-2 if applicable.

Any such hose needs to be installed in a manner that does not result in any strain or torsion and 90° action manual valve will be required upstream of the flexible pipe assembly.

Ventilation

Adequate ventilation must be provided to permit the safe operation of the appliance(s). BS 6896, BS EN 13410 and the commercial ventilation section of the ConstructionSkills reference manual (G23 or G88) gives detailed information on calculating air supply requirements for any given radiant heating application.

Means of ventilation for unflued appliances will be required by the following methods:

- thermal evacuation – convective evacuation of the products of combustion/air mixture through defined openings in the roof or the walls of the building

- mechanical evacuation – evacuation of the products of combustion/air mixture by fans in the roof or the walls of the building

- natural air change – evacuation of the products of combustion/air mixture through undefined openings by pressure differences and temperature differences between the inside and outside of a building.

Where natural ventilation is provided, grilles need to be located at low level for flued applications and at both low and high level for unflued applications. Unlike floor standing boilers and warm air units however, low-level ventilation openings for radiant applications need to be sited below the heater(s), not necessarily within 1 m of floor level. Openings at high level are to be sited above the heater(s). Vertical distance between high and low openings should not be less than 3 m.

Ventilation openings need to be sited where they cannot easily be blocked or flooded, ideally on at least two sides of the building and located so that suction of the exhaust air is not disturbed by wind influence.

When considering exhaust air openings the airflow rate needs to accommodate requirements for other purposes. The size and number of openings will be based on the higher of these airflow rates.

Flues

Requirements for flue systems are to be found in BS 6896 and the commercial flues section of the ConstructionSkills reference manual.

Any flue system needs to be constructed of non-combustible materials that are not unduly affected by heat, condensation and the products of combustion.

Manufacturer's information should be consulted to ensure that maximum lengths of flue pipe are not exceeded. The resistance of the flue system must not be greater than the capability of the exhaust fan.

Where any particular appliance is not rigidly supported, i.e. tube heaters, corrugated flexible flue connections need to be fitted. The maximum length of flexible connection should not exceed 1 m.

Also, of particular importance for some radiant tube systems is the need to ensure that the flue can be easily disconnected to facilitate correct maintenance procedures and repairs.

WATER BOILERS AND SYSTEMS

Water boilers are often used for the indirect fired heating of a given space, whereby the heated water is circulated through a piped system connected to of a number of radiator panels or convector heaters. However, in commercial or industrial applications they may also be used to provide hot water for particular processes.

Water boiler systems for the industrial and commercial utilisation market come in a wide variety of shapes, sizes and arrangements from little more than a domestic boiler up to cast iron section or steel 'shell' type boilers having output capacities of 6 MW.

Types of boiler systems include the following:

- natural draught boilers, conventionally flued or balanced flue
- fanned draught boilers, conventionally flued or balanced flue
- fan dilution boiler systems
- modular boiler systems
  - natural draught or fanned draught
  - low water content, high efficiency
- condensing boilers
- storage water heaters
- cast iron section boilers
- steel shell boilers.

NATURAL DRAUGHT (CONVENTIONAL 'OPEN' FLUE)

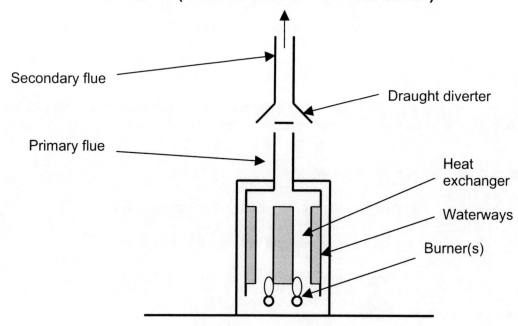

Schemmatic view conventional flue boiler

Older boiler models are traditionally floor standing and constructed of cast iron. More modern types are likely to be constructed of finned tubes with lower water content. These modern type boilers will therefore be lighter in weight and smaller output models can sometimes be wall mounted rather than floor standing as a consequence.

Typically these types of boiler will comprise a partial aerated burner sited beneath the heat exchanger which entrains roughly 50–60% of combustion air at primary air ports at the injector. The remaining air requirement for combustion is secondary air taken from around the point of combustion. Combustion takes place in the combustion chamber beneath the heat exchanger and the products of combustion then pass upward through the heat exchanger transferring heat to the water medium.

There should be no flame contact with the heat exchanger or the relatively cold metal surfaces will quench the flame causing 'flame chilling'. This will result in loss of boiler efficiency but more importantly may cause formation of carbon monoxide due to the interruption of the chemical reaction when combustion takes place.

Natural draught systems rely on the natural state of hot gases rising through the heat exchanger and into the flue exhaust. This upward movement will naturally pull replacement air into the combustion chamber through the air inlets on the boiler assembly.

The flue pipe will normally include a break or 'draught diverter' at or near its connection to the boiler. This is designed to assist the natural pull of the flue by sucking in additional air, diluting the combustion products. It will also prevent any prevailing wind affecting combustion by diverting down-draught away from the point of combustion. Without this facility adverse conditions in the flue would prevent the natural path of the rising combustion gases to free air, stifling the flame, creating the condition of incomplete combustion and the production of carbon monoxide.

NATURAL DRAUGHT (BALANCED FLUE)

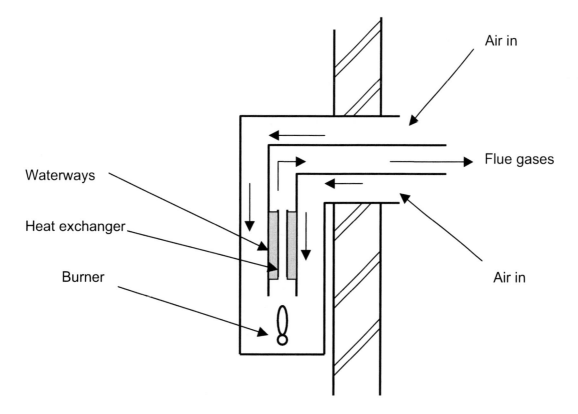

Schemmatic view wall mounted balanced flue boiler

Balanced flue boilers are generally available up to approximately 50 kW, although some particular units are available up to a nominal 100 kW capacity.

They are effectively a 'sealed' appliance. Fresh air for combustion is drawn in from external atmosphere, unlike open flued appliances where combustion air is taken from within the internal room. This will have the benefit of minimising the need for additional ventilation air bricks/grilles to external atmosphere (which may give rise to draughts of cold air).

Other advantages of fitting a balanced flue boiler include:

- no need for tall external unsightly flue pipe

- reduced installation costs (no additional flue pipe required)

- air inlet and flue outlet are immediately adjacent therefore any variations in wind pressure affect both equally

- packaged unit.

The main disadvantage of a balanced flue appliance is the requirement for it to be located adjacent to an outside wall, this can present restrictions for suitable boiler locations. Also, semi-condensing or condensing boilers, in cold weather particularly, can produce significant volumes of steam at the terminal. Consideration should therefore be given to building structure. Solid walls may result in damp conditions internally.

FAN DRAUGHT (FORCED OR INDUCED – CLOSED FLUE)

When faced with location restrictions on conventional or balanced flue boilers due to impractical flue pipe route or lack of outside wall, it may be possible to install a fan assisted flue system.

This will enable the installer to install some horizontal flue pipe (subject to manufacturer's instructions) from the appliance to external atmosphere and still allow adequate dispersal of combustion gases. If, however, an induced draught fan is fitted in the flue it must obviously be capable of withstanding the high temperature of moist exhaust gases.

FAN UPSTREAM OF THE BURNER – FORCED DRAUGHT

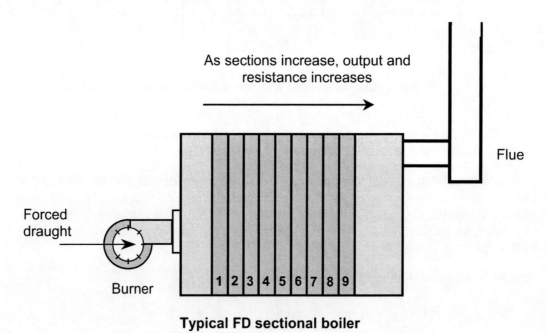

Typical FD sectional boiler

Traditionally a forced draught burner will be used on larger capacity boilers nominally from 35 kW and above on floor standing cast iron section or steel boilers.

Typically these types of boiler will utilise a 'packaged' nozzle mix burner sited at the front, firing horizontally through a combustion chamber. The combustion chamber is located beneath the heat exchanger so that second (and third) passes of the combustion gases transfer heat through the boiler walls to the water medium.

Again, there should be no flame contact with the combustion chamber (and heat exchanger) or the relatively cold metal surfaces will quench the flame causing 'flame chilling'. As with natural draught boilers, this will result in loss of boiler efficiency and may cause formation of carbon monoxide.

Back end boiler protection, such as target brick, will be required to prevent heat damage from the gas flame.

The nozzle mix burner allows the entire air requirement for combustion to be mixed with the fuel gas at the burner head. There is no need to supply a source of secondary air at the point of combustion as with natural draught boilers.

Also, because the air/gas mixture is being forced into the combustion chamber, heat exchanger and flue system, there is less reliance for the natural buoyancy of the flue system to 'pull' the products of combustion through the boiler to the point of exit.

In consequence, a forced draught boiler will be more compact size for size when compared to a natural draught boiler of similar heat output capacity.

The flue pipe may include a 'draught stabiliser' at or near its connection to the boiler. A draught stabiliser is a mechanical weight loaded damper mechanism, which is designed to prevent too much suction from the flue being applied to the burner. Too much suction in the flue causes the damper to open. Adjustments for the correct operation of the damper mechanism will be given in manufacturer's information.

FAN DOWNSTREAM OF THE BURNER – INDUCED DRAUGHT

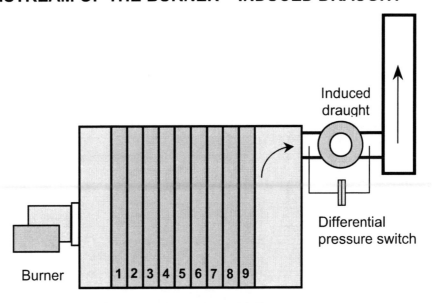

Typical ID sectional boiler

Induced draught burner systems are becoming increasingly popular on smaller capacity and domestic boilers particularly. This is part due to the fact that modern natural draught boilers are more efficient, generating less heat to the flue, and therefore less buoyancy to pull the products of combustion upwards through the flue system. As a consequence some kind of fan assistance is required to disperse to the combustion products correctly.

Typically these types of boiler will utilise a 'post-aerated' gas burner, which comprises a gas nozzle and air diffuser plate. The induced draught fan will 'suck' air in through the air holes in the diffuser plate. The size and configuration of the air holes being graduated to give correct air/gas mixing.

FAN DILUTION SYSTEMS

The fan diluted flue system was developed to overcome difficulties which arise when running long lengths of vertical flue pipe to disperse the products of combustion from the boilers above buildings, i.e. high rise buildings or ground floor premises with offices over the boiler plant room, etc.

Apart from the aesthetically displeasing view of high rise flue pipes running externally up a prestigious building (hotels, prime office accommodation, etc.) there may also be installation cost benefits of fitting a fan dilution system instead of large diameter flue pipe.

The principle relies on an exhaust fan fitted in the flue outlet from the boiler(s) which pulls in large amounts of fresh air to dilute the products of combustion to below 1% CO_2. Products of combustion can then be discharged to atmosphere at low level.

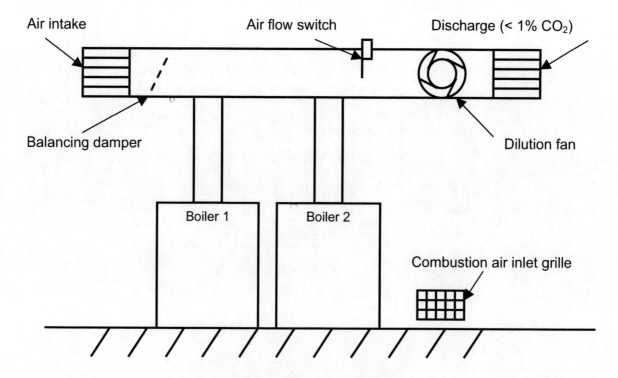

Typical fan dilution flue system

Forced or natural draught appliances are connected to a header duct by vertical flue pipe. At one end of the header duct is the dilution air intake and at the other the discharge grille. A fan is located at the discharge end to draw in the fresh dilution air, which is then mixed with the combustion gases, diluted and discharged at low level.

Discharge height for boilers rated up to 1 MW can be as low as 2 m above external ground level. Above this capacity the discharge height is a minimum of 3 m.

Other important features of the system are:

- air flow proving device interlocked with the boilers to prevent firing in the event of air flow failure

- balancing damper to control dilution air throughput

- inlet and outlet ducts should be sited on the same wall if possible to balance wind effect

- exhaust gases must not discharge to an enclosed area, i.e. courtyard

- discharge outlets must be directed at an angle above the horizontal (30°)

- discharge grilles must be positioned away from openable windows and adjacent buildings – IGE/UP/10, the 1956 Clean Air Act Memorandum and BS 6644, give advice on minimum spacing.

MODULAR BOILER SYSTEMS

Modular boiler systems usually comprise of three or more identical natural draught boiler 'modules', interconnected to provide incremental heat build up to match heat demand, particularly where there are significant differences between peak demand and normal running. The number of boilers operating at any time is only that required to meet the heat demand.

A step controller is used to bring on each additional boiler as the heating load increases and shut them down again as the load decreases. Some controllers will cycle the lighting sequence of the boilers to optimise the use (and subsequently wear) of all boilers, rather than continually start the heat up sequence with the same boiler.

If a single capacity boiler were to be used in this scenario, there would be significant wasted energy in keeping a large boiler and its water content heated at periods of low heat demand.

The preferred flue arrangement for any modular system would be to independently flue each boiler. However, due to cost implications and the impracticality of individual flue systems, it is usual to find each boiler flued to a common header system.

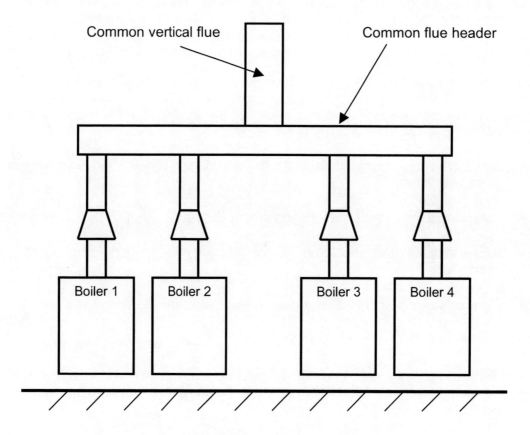

Modular boiler system

However, there are a few design requirements of such a system.

- common flue must be larger than the individual flues to cope with combined flow rates when all boiler modules are firing

- maximum number or six boiler modules to a common horizontal flue

- maximum number or eight boiler modules to a common vertical flue

- connector flue between each boiler module and the common header should not be less than 0.5 m from the draught diverter base

- the connection between the individual flue at the common header should be a smooth bend or sloping 135° connection.

HIGH EFFICIENCY MODULAR BOILER SYSTEMS

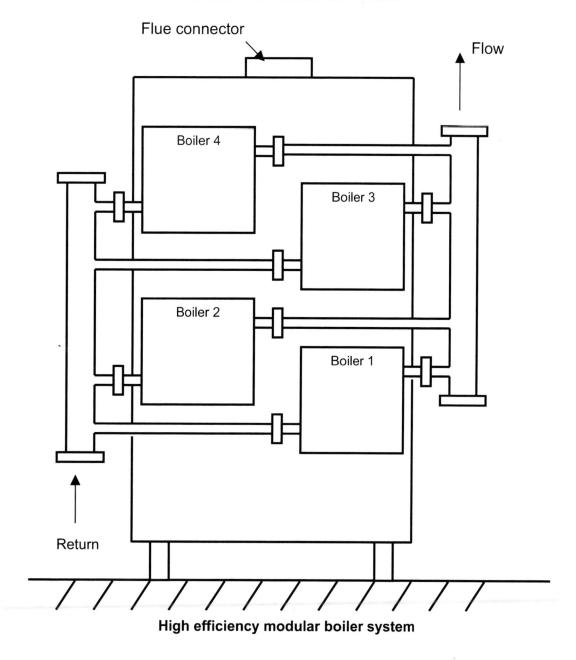

High efficiency modular boiler system

This variation of modular boiler system offers a compact system of standard capacity boilers (usually 50 kW or 100 kW) which are added together to provide the heat capacity required for a given maximum load.

With some particular manufacturer's systems, the boilers are controlled such that, as the load is satisfied the uppermost boiler is switched off first and then in descending boiler order. This arrangement permits the water in the upper boiler modules to be heated by the exhaust gases from the lower module.

Apart from the obvious energy saving from this system, and the low water content, these units offer considerable space saving; in most cases less than a third of the floor space required for conventional boilers.

CONDENSING BOILERS

By recovering the latent heat from the water vapour in combustion gases, additional energy savings can be made from any heating system.

Generally the heat in the flue gases leaving a conventional boiler will be approximately 250°C. To recover the latent heat from the water vapour in the gases it is necessary to drop the flue gas temperature down to below its dew point, approximately 55°C.

The essential difference between a conventional boiler and a condensing boiler is the addition of a secondary heat exchanger in the latter.

The secondary heat exchanger needs to be constructed using stainless steel or aluminium alloy in order to protect it from the effects of the mildly acidic condensate.

An exhaust fan is fitted to draw the products of combustion through the secondary heat exchanger.

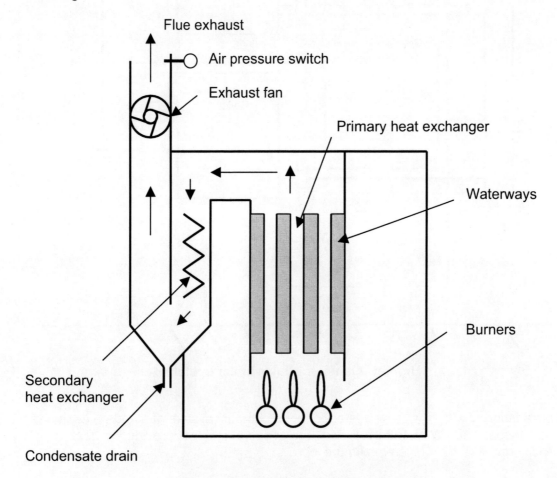

Schemmatic view of condensing boiler

STORAGE WATER HEATERS

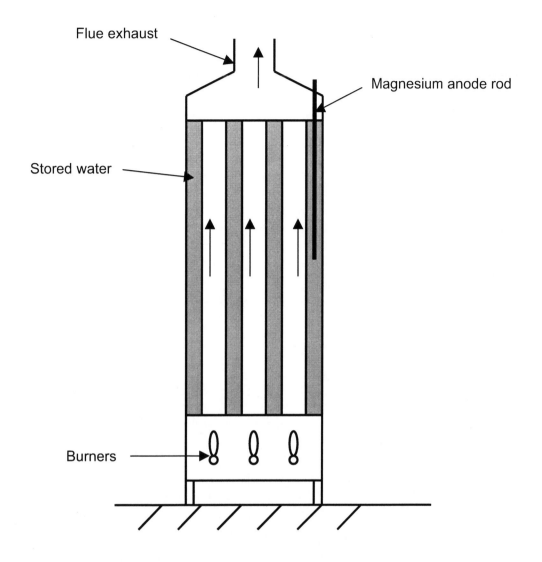

Schemmatic view of typical storage water heater

Direct-fired storage water heaters are usually natural draught appliances, which comprise a cylinder containing water with a number of heat exchanger tubes passing through. The cylinder is under-fired such that the heat from the combustion gases passes through to the water contained in the cylinder. However, increasingly forced draught burners are available.

The hot water can then be drawn from the cylinder directly and because the water is heated directly, i.e. no calorifier, it is a more efficient system.

To avoid problems of Legionnaires disease, associated with storage of warm water, stored water temperature should be kept above 65°C. In addition, it has become common practice to install a pumped re-circulation system to ensure even water temperature.

Although the cylinder can be glass lined for protection, damage can sometimes occur. On some appliances protection is provided by means of an anode, whilst others may use a small impressed current in the water. In either case the protection is provided to minimise boiler corrosion. New units are increasingly made from stainless steel.

Scaling of the boiler can sometimes be a particular problem associated with this type of appliance. Manufacturers usually recommend use of a chemical descaling agent at regular intervals.

CAST IRON SECTIONAL BOILERS

Sectional boilers are in many ways the traditional type of boiler and comprise a number of standard cast iron section pieces joined together. They offer a degree of versatility due to the fact that to increase output capacity to hot water, all that the boiler manufacturer needs to do is add extra sections. Any standard model of boiler can therefore be offered in a wide range of output capacities.

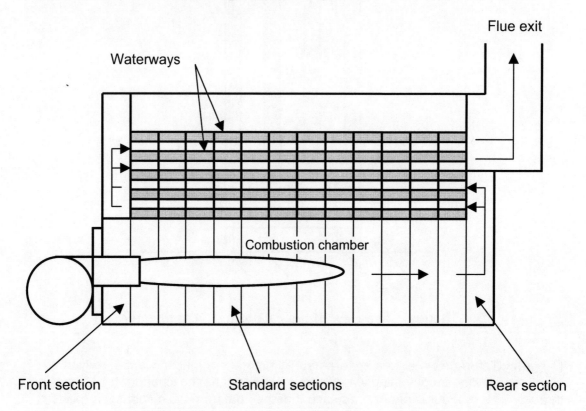

Schemmatic view of large cast iron sectional boiler

The above illustration indicates the typical layout of a horizontal forced draught fired boiler with sections assembled horizontally.

In the forced draught fired example above, the products of combustion are forced from the combustion chamber into the horizontal heat exchanger with two passes through to the flue exit. Heat is transferred to the waterways via the heat exchanger.

For natural draught fired boilers the boiler sections could be assembled horizontally or vertically, dependent on the manufacturer's cast form of section. Natural draught sectional boilers must allow products of combustion to rise through the boiler, transferring the heat through the heat exchanger to the water.

In principle if a section fails for any reason it may be possible, dependent upon age and general condition of the boiler, to replace the individual section rather than the complete boiler.

STEEL SHELL BOILERS

Shell boilers are a fabricated, cylindrical, welded steel construction built to provide a nominal individual heat capacity output within a total range, in general, above 50 kW and up to 6 MW. They are also designed for high-pressure hot water and steam generating applications.

Because they are built to withstand higher operating pressures they are of a more compact design. They can be up to 50% smaller than the cast iron sectional boiler of similar output capacities.

However, this will generally mean that the higher firing resistance through the combustion chamber and subsequent passage through the heat exchanger requires higher gas firing pressures than would normally be supplied through the local distribution system. If elevated gas pressure from the distribution system is not available it will usually require a local gas booster to lift incoming pressures to meet boiler manufacturer's recommendations.

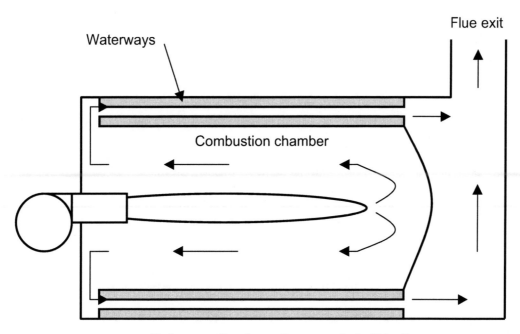

Schemmatic view of reversal shell boiler

The above illustration indicates the typical layout of a 'Reversal' steel shell boiler where the products of combustion pass back along the path of the flame to the front of the boiler before entering the heat exchanger tubes.

INSTALLATION REQUIREMENTS

Water boilers must be installed in accordance with the relevant provisions of the Gas Safety (Installation & Use) Regulations and obligations that arise as a result of the Health and Safety at Works Act.

When installing gas fired heating systems, guidance should be sought from appropriate reference standards. The <u>specific</u> relevant reference standards are as follows:

BS 5440: 2000   Specification for installation and maintenance of flues and ventilation for gas appliances of rated input not exceeding 70 kW net (1st, 2nd and 3rd family gases). (N.B. 2nd family gases = natural gas)

BS 6644: 2005   Specification for installation of gas–fired hot water boilers of rated net inputs between 70 kW and 1.8 MW (2nd family gases)

IGE/UP/10 Edition 3   Installation of Gas Appliances in Industrial and Commercial Premises

MANUFACTURER'S INFORMATION

Manufacturer's recommendations should always be consulted prior to installation work commencing in order to establish all particular requirements, but these are likely to cover the following criteria.

Location

Any appliance needs to be located in a safe and secure manner where it will not be damaged by other activities.

Many new buildings, such as office suites or hotels, locate the heating plant in roof top installations, although more traditional commercial buildings may locate the plant in basements or ground floor areas.

Due consideration will need to be given to suitable protection for any boiler sited in basement areas that may be subject to flooding. High level boiler installations, on the other hand, will require consideration for protection against low system water level or frost. Precautions may also be required to protect the building against the effects of water leakage from a roof top installation.

Boiler rooms must be carefully designed or selected and ensure that no point within that boiler room is more than 12 m from the nearest means of access.

Typically boilers will be floor mounted but some manufacturer's boilers may be fixed on purpose made brackets from a wall.

Any means of support must be of sufficient number and strength to support the full weight of the boiler (when filled with water) and the unsupported part of its flue system.

Floor standing boilers need to be stable or rigidly fixed.

The floor on which any boiler may be sited must not be subjected to temperatures in excess of 65°C when the boiler is in operation. Adequate clearance must be provided between any combustible materials, or any material likely to be affected by high temperatures, and the appliance or flue. (Manufacturer's information will usually provide guidance on minimum clearances.) The boiler room must not exceed set temperatures at different levels. Please see the section on ventilation for clarification.

Due consideration must also be given to heat emission radiated from the boiler installation to ensure that the boiler room temperature and that of surrounding internal rooms remain compatible during the boiler operation with their intended uses.

As with any gas appliance in the commercial and industrial environment, consideration must be given to the location of the boilers and the atmosphere where they are to be used. Boilers must not be installed within buildings that have been classified as hazardous areas. The definitions for hazardous areas can be found in BS EN 60079-10.

Boilers supplying heat to hazardous areas need to be located either in a separate room/compartment or in the open air. If the boiler is sited in a compartment adjacent to the area where flammable, heavier than air vapours may be present, the compartment itself will be subject to certain requirements. For example, a 450 mm high non-combustible doorsill, self closing door and one hour fire resistance, etc. Other conditions may be applicable which affect the boiler position, such as location of fresh air intake.

Siting a boiler in garage workshops may present problems due to the presence of petroleum and other heavier than air vapours. In this environment the boiler controls (time switches, thermostats) must not be sited within 1.2 m of the floor level within the space to be heated. The boilers themselves must be sited at least 1.8 m above the floor level.

Any boiler must be suitable for the conditions in which it is to be installed, particularly if sited externally. Alternative materials for finishes and manufacture may be necessary in corrosive environments. This also applies to the installation materials such as gas pipe.

Gas Pipework

The general pipework installation should comply with the 'Institution of Gas Engineers' reference standard UP/2 – Gas Installation pipework, boosters and compressors on industrial and commercial premises.

The gas supply pipework needs to be adequately sized to provide the working pressure for the volume required by the boiler(s) to maintain the nominal heat input.

The Gas Safety (Installation & Use) Regulations requires that a suitable isolation valve be fitted to the appliance. A pipe union will usually be required downstream of the appliance isolation valve to permit burner removal for maintenance purposes.

Plant pipework upstream of the burner needs to be adequately supported, making allowance for any vibration and/or movement. There may be a requirement to connect the boiler to the gas supply system using a corrugated metallic gas flex manufactured to BS 6501-1 or 669-2, for example, if a gas booster is installed. Manufacturer's advice should be sought.

Where such a flex is required it needs to be installed in a manner that does not result in any strain or torsion and a 90° action manual valve will be required upstream of the flexible pipe assembly.

Ventilation

Adequate ventilation must be provided to permit the safe operation of the appliance(s).

BS 5440, BS 6644 and the commercial ventilation section of the ConstructionSkills reference manual gives detailed information on calculating air supply requirements for any given heating application.

Where natural ventilation is provided, grilles need to be located at low level (below the appliance flue connection).

Ventilation openings need to be sited where they cannot easily be blocked or flooded, ideally on at least two sides of the building, and located so that suction of the exhaust air is not disturbed by wind influence.

When considering exhaust air openings the airflow rate needs to accommodate requirements for other purposes. The size and number of openings will be based on the higher of these airflow rates.

Flues

Requirements for flue systems are to be found in BS 5440, BS 6644, IGE/UP/10, manufacturer's instructions and the commercial flues section of the ConstructionSkills reference manual. For boilers exceeding 150 kW the 3rd edition of the 1956 Clean Air Act Memorandum also applies regarding flue termination and minimum flue gas velocity.

Any flue system needs to be constructed of non-combustible materials that are not unduly affected by heat, condensation and the products of combustion.

Manufacturer's information should be consulted to ensure that maximum lengths of flue pipe are not exceeded. Where exhaust fans are fitted, the resistance of the flue system must not be greater than the capability of the exhaust fan.

Also of particular importance for some types of boiler is the need to ensure that the flue can be easily disconnected to facilitate correct maintenance procedures and repairs.

An approved type terminal will be required on flue pipe systems up to 170 mm diameter, but above this may not be necessary.

The cross sectional area of an open flue (natural draught) needs to be at least the same size as the flue outlet spigot of the boiler. However, on induced draught systems the flue size may be reduced, subject to manufacturer's instructions.

Any fan assisted means of exhausting flue products needs to incorporate a protection device to shut down the boiler in the event of air flow failure.

When installing common flue systems for natural draught modular boiler installations there are a few basic ground rules:

- avoid restrictions (sharp changes in flue direction, abrupt transitions)
- flue headers should be installed as high as the boiler room will permit
- flue headers should be a constant size
- connector sections from the draught diverter base to the header should be at least 0.5 m high
- shared boilers should be fitted in the same area
- appliances connected to a common flue must be of the same burner type, for example, natural or forced draught.

Other types of flue/ventilation systems include balanced compartments. These are of particular benefit for boiler rooms with no adjacent outside wall, which would otherwise require a long ducted ventilation system.

Balanced Compartment or High Level Supply Discharge

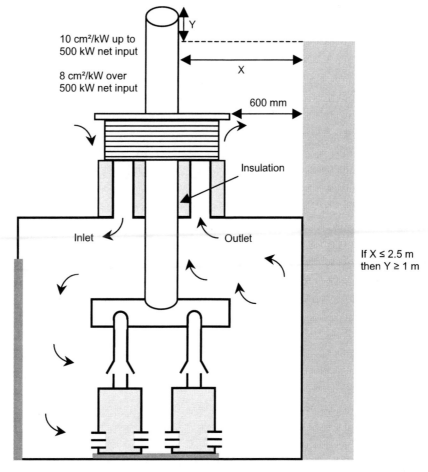

Typical balanced compartment

A specialist area of design and installation, balanced compartments for non-domestic applications differ to that of domestic appliances.

The environment of the compartment must be under controlled conditions. All surfaces, pipework, flues or ducts that may affect air movement, heat transfer or dissipation have to be adequately insulated.

Any door is self-closing and ensures a seal for the compartment. Doors are usually fitted with electrical switches that will shut down the appliances if the door is opened. The door requires a notice warning that it should not be opened unless for control adjustment or maintenance.

Ancillary Equipment

Installations require water temperature gauges to be located on the flow from the boiler.

Safety valves can depend on the design pressure and input rating of the boiler(s).

For installations of more than one boiler, modular installations require a common safety valve unless each boiler is already fitted with a safety valve complying with BS 6759-1. Multiple individual boiler installations will require a suitable safety valve on each boiler.

Discharge from any safety valve needs to be self-draining and terminate in a visible position. Any such discharge must not result in a hazard to personnel or plant.

Every boiler connected in parallel to a common flow and return manifold to form a multiple boiler installation needs to be fitted with individual means of isolation from the water and electricity as well as gas supplies. Individual boilers in modular boiler installations, however, need not be provided with water isolating valves provided they are fitted on the common flow and return pipes of each bank of boilers.

COMMISSIONING GAS EQUIPMENT

Traditionally the term 'commission' is used to describe the final testing and operational checks carried out at the stage directly after an appliance has been installed but has yet to be brought into service.

An engineer has various sources from which to construct a commissioning procedure for an appliance or indeed larger scale equipment.

- The Gas Safety (Installation and Use) Regulations, Regulation 26 and 33.
- Manufacturer's instructions.

And, in the absence of concise or current manufacturer's instructions:

- British Standards documents.
- Institution of Gas Engineers, publication IGE/UP/4.

Each of the above sources offers instruction as to what should be addressed in the commissioning of appliances or plant. The manufacturer's instructions should be the most specific in relation to an appliance, with the manufacturer being the expert. This may not always be the case and sometimes manufacturer's instructions or British Standard documents can be dated and even neglect certain areas of gas safety that are common practice in today's environment.

To ensure a full, concise and complete commissioning procedure is adopted for each case, an engineer needs to add two more ingredients to the list above:

- sound engineering judgement, and
- good working practice.

Depending on the nature of the installation extensive planning, communication with other parties, design verification and programming may be required before installation and commissioning.

Formal programming may consist of the following:

Planning and Programming Period

This period can include the collection of documentation and verification of plant design.

The formal commissioning procedure is created and agreed with all relevant parties, which may include the manufacturer, the senior/responsible engineer, the commissioning engineer and site or plant engineer.

The programming of works should include the input of the customer/client, consultants, all main and sub contractors, insurers and fuel suppliers and the issue of a 'Permit to Work' where necessary.

Inspection Period

The inspection of all systems and components before services are introduced.

Activation Period

Usually termed as the 'Dry Run' and 'Live Run', this period allows an engineer to identify operational faults or safety concerns before operating levels are established.

The 'Dry Run' should include the checking of controls and interlocks including, process controls and fault condition systems where necessary.

Operation Period

Establishing the satisfactory performance of all systems, controls and ensuring correct operation to appropriate settings, this should include combustion analysis and recognised acceptance of output levels.

Completion Period

Also known as the 'Hand Over' period, includes instruction to the client and relevant parties that may be required to operate the equipment in operating, shut down and starting procedures.

It is important to ensure that relevant parties are aware of the necessary action to be taken in the event of problems, faulty operation, shut down and emergency condition.

When commissioning is complete instruction to the users and hand over procedures need to be concise and appropriate for the expectations of the user.

The following generic procedural framework intends to suggest areas for attention when commissioning a commercial gas appliance. It is not intended to be concise or complete and should only be used as a tool within a training environment, in compiling a commissioning procedure in addition to the manufacturer's instructions.

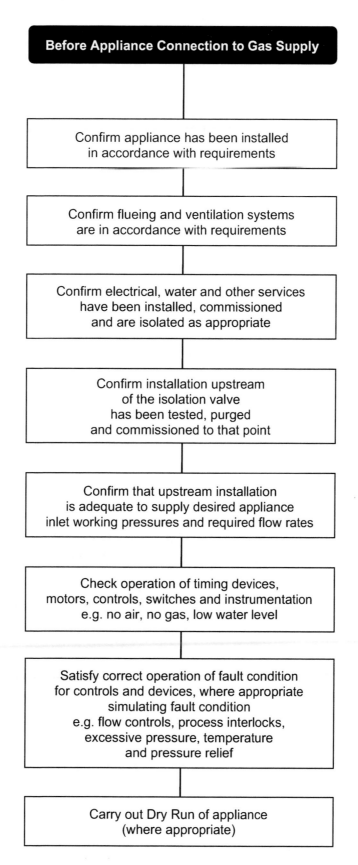

Typical commissioning procedural framework

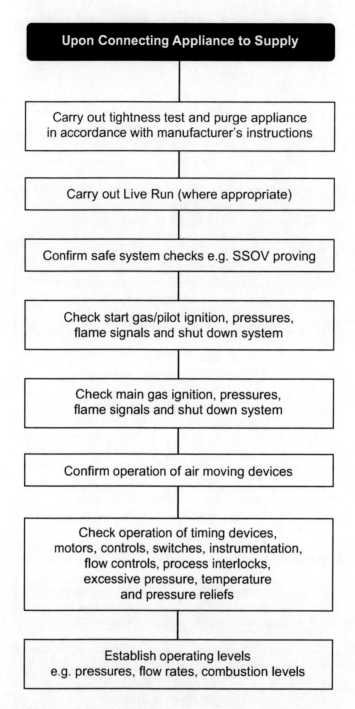

Typical commissioning procedural framework

The following typical procedure aims to show the order in which many operations may be carried out.

Inspection Period

- The gas supply is positively isolated.

- The gas supply is confirmed as correct and suitable, e.g. supply and pressure.

- The upstream supply is confirmed as tested, purged and commissioned by documentation.

- The electrical supply is isolated.

- Electrical cables confirmed not to be damaged when the equipment/plant is in operation.

- Any hydraulic or pneumatic supplies are positively isolated.

- The equipment/plant and controls are visually inspected against the specification.

- Emergency isolation valves are confirmed as operating correctly and clearly identified.

- Adequate testing and purge points are confirmed as available.

- The ventilation and flueing requirements for the equipment/plant are checked and confirmed as correct, taking into consideration the presence of other appliances/equipment/plant.

- All earthing and bonding is checked and confirmed as correct.

- No leakages or spillages of oil, solvent or water are present.

- Adjacent machinery or plant is confirmed not to cause a hazard whilst commissioning.

- Appropriate tools are chosen and available.

- Testing equipment and instrumentation is calibrated and available.

- Associated controls and equipment are ready for use.

- Appropriate safety systems are operative.

Dry Run

- The equipment/plant is tested for gas tightness in accordance with industry guidance and the manufacturer's instructions.

- Safety shut off valves (SSOVs) and non return valves are confirmed/proven in the closed position.

- The equipment/plant is purged in accordance with industry guidance and the manufacturer's instructions.

- Interlocking devices/controls are set to provisional operating levels, considered safe, e.g. dampers, governors and regulators, pressure reliefs, position, pressure and flow control systems and switches.

- Electrical equipment/controls/interlocks are checked for correct sequencing and operation ensuring, e.g. (when sanctioned by the manufacturer, flame simulators may be used and interlocks may be 'linked out'):
  - the combustion area/space is adequately purged (flow rate and timing) before checking ignition source
  - timing devices, dampers, interlocks, motors, flow controls and systems operate satisfactorily
  - valves are checked and valve proving systems operate correctly
  - safe start check functions are proven for at least two consecutive operations (flame safeguards)
  - main flame ignition air flow rate is correct
  - air flow rates are checked and ignition sources proven operational under ignition air flow rate conditions
  - a simulated flame signal is recognised and lock out/shut down proven to occur within time requirements
  - the sequencing is proven as correct, e.g. pre-purge, valve proving, ignition, flame proving, opening and shutting of valves
  - safety shut off valves are proven 'leak tight' after operation
  - cooling medium is supplied as required (e.g. UV detector heads)
  - the shut down sequence is proven as correct (including post-purge where applicable).

- All interlocks are reinstated.

Live Run

- Main gas is prevented from flowing to the burner.
- Start gas check:
  - combustion area/space is correctly purged
  - dampers/controls are correctly set before ignition
  - stable start/pilot flame is confirmed
  - pipework downstream of any start/pilot shut-off device is confirmed as 'gas tight' (where it cannot be tested or confirmed under a tightness test)
  - flame signal is confirmed as satisfactory
  - shut down occurs correctly when start/pilot gas is extinguished
  - main line shut-off valves remain closed
  - sequence is re-checked for start/pilot ignition/shutdown and the failure of main ignition.
- Main gas is turned on.
- Main gas check:
  - combustion area/space is correctly purged
  - dampers/controls are correctly set before ignition
  - stable start/pilot flame is established
  - stable main flame is confirmed and established
  - pipework downstream of any main shut-off device is confirmed as 'gas tight' (where it cannot be tested or confirmed under a tightness test)
  - flame signal is confirmed as satisfactory
  - shut down occurs correctly when gas is extinguished
  - sequence is re-checked for ignition/shutdown and the failure of main flame
  - interlocks are proven to operate correctly.
- Operational checks:
  - air/gas ratios are set in accordance with manufacturer's instructions
  - flames are confirmed to be stable across varying gas rates
  - combustion levels and efficiency are checked in accordance with manufacturer's instructions
  - flue stack temperatures and flue gas analysis efficiency are checked in accordance with manufacturer's instructions
  - remaining controls and interlocks are checked for correct operation
  - upon shut down, safety shut off valves are confirmed as 'closed'.
- Completion of report and documentation.

Commissioning

The following gives a typical procedural framework for the commissioning of a **radiant plaque heater**.

- Ensure the gas supply is of adequate size for the appliance.

- Confirm the appliance is sited in accordance with manufacturer's instructions, e.g. correctly suspended at a suitable height in accordance with manufacturer's instructions.

- Confirm the appliance assembly is complete and fit for intended use and purpose.

- Confirm gas pipework, fittings, isolation valves, controls and flexible metallic hoses conform to requirements.

- Isolate gas/electrical supplies as appropriate prior to work being carried out.

- Carry out gas tightness testing, purging and electrical testing as appropriate.

- Re-establish gas/electrical supplies as appropriate.

- Record and confirm the working pressure to the appliance is correct.

- Record and confirm the appliance operating pressure (adjusted as necessary).

- Ensure burner ignition is correct and flame pictures are stable.

- Record and confirm the heat input of the appliance.

- Confirm ventilation requirements are correct and in accordance with manufacturer's instructions.

- Ensure safety control devices are operating correctly.

- Carry out a CO_2 atmosphere test and record results.

- Confirm user controls are operating correctly, e.g. thermostat (if applicable).

- Identify and record any defects or faults and take necessary action.

- explain the safe operation and use of the appliance to responsible persons and users.

The following gives a typical procedural framework for the commissioning of a **radiant tube heater**.

- Ensure the gas supply is of adequate size for the appliance.

- Confirm the appliance is sited in accordance with manufacturer's instructions, e.g. correctly suspended at a suitable height in accordance with manufacturer's instructions.

- Confirm the appliance assembly is complete and fit for intended use and purpose.

- Confirm gas pipework, fittings, isolation valves, controls and flexible metallic hoses conform to requirements.

- Confirm correct installation, connection and use of materials for any flue serving the appliance.

- Isolate gas/electrical supplies as appropriate prior to work being carried out.

- Carry out gas tightness testing, purging and electrical testing as appropriate.

- Re-establish gas/electrical supplies as appropriate.

- Record and confirm the working pressure to the appliance is correct.

- Record and confirm the appliance operating pressure (adjusted as necessary).

- Ensure burner ignition is correct and flame pictures are stable.

- Record and confirm the heat input of the appliance.

- Confirm ventilation requirements are correct and in accordance with manufacturer's instructions.

- Ensure safety control devices are operating correctly.

- Carry out appropriate flue testing to manufacturer's instructions, including combustion analysis.

- Carry out a CO_2 atmosphere test and record results if unflued.

- Confirm user controls are operating correctly, e.g. thermostat (if applicable).

- Identify and record any defects or faults and take necessary action.

- Explain the safe operation and use of the appliance to responsible persons and users.

The following gives a typical procedural framework for the commissioning of an **indirect fired air heater**.

- Ensure the gas supply is of adequate size for the appliance.
- Confirm the appliance is sited and supported in accordance with manufacturer's instructions.
- Confirm the appliance assembly is complete and fit for intended use and purpose.
- Confirm the correct installation and suitability of any ductwork and flue.
- Confirm gas pipework, fittings, isolation valves, controls and flexible metallic hoses conform to requirements.
- Confirm appliance is correctly secured to ductwork (if applicable).
- Isolate gas/electrical supplies as appropriate prior to work being carried out.
- Carry out gas tightness testing, purging and electrical testing as appropriate.
- Re-establish gas/electrical supplies as appropriate.
- Record and confirm the working pressure to the appliance is correct.
- Record and confirm the appliance operating pressure (adjusted as necessary).
- Ensure burner ignition is correct and flame pictures are stable.
- Record and confirm the heat input of the appliance.
- Confirm ventilation requirements are correct and in accordance with manufacturer's instructions.
- Set combustion ratio in accordance with manufacturer's instructions (forced draught burner), including combustion analysis.
- Test flue system appropriately and confirmed to be correctly clearing products of combustion.
- Ensure the heat exchanger is not leaking by procedure given by manufacturer.
- Ensure safety control devices are operating correctly, including high limit and fan thermostats.
- Confirm user controls are operating correctly, e.g. thermostat (if applicable).
- Identify and record any defects or faults and take necessary action.
- Explain the safe operation and use of the appliance to responsible persons and users.

When commissioning a direct fired air heater, in addition to the relevant aforementioned tasks, a CO_2 atmosphere test should be carried out and the results recorded.

The following gives a typical procedural framework for the commissioning of an **indirect water heater/boiler**.

- Ensure the gas supply is of adequate size for the appliance.

- Confirm the appliance is sited and supported in accordance with manufacturer's instructions.

- Confirm the appliance assembly is completed and fit for intended use and purpose.

- Confirm the correct installation and suitability of the flue system.

- Confirm gas pipework, fittings, isolation valves, controls and flexible metallic hoses conform to requirements.

- Isolate gas/electrical supplies as appropriate prior to work being carried out.

- Carry out gas tightness testing, purging and electrical testing as appropriate.

- Re-establish gas/electrical supplies as appropriate.

- Record and confirm the working pressure to the appliance is correct.

- Record and confirm the appliance operating pressure (adjusted as necessary).

- Ensure burner ignition is correct and flame pictures are stable.

- Record and confirm the heat input of the appliance.

- Confirm ventilation requirements are correct and in accordance with manufacturer's instructions.

- Set combustion ratio in accordance with manufacturer's instructions (forced draught burner), including combustion analysis.

- Test flue system appropriately and confirmed to be correctly clearing products of combustion.

- Ensure the heat exchanger is not leaking by procedure given by manufacturer.

- Ensure safety control devices are operating correctly, including high limit and fan thermostats.

- Confirm user controls are operating correctly, e.g. thermostat (if applicable).

- Identify and record any defects or faults and take necessary action.

- Explain the safe operation and use of the appliance to responsible persons and users.

SERVICING

Regular servicing of gas appliances, plant and equipment is essential to maintain their optimum performance and ensure that all fitted safety features are always effective.

In addition, particularly with some older appliances/plant/equipment, it may be necessary to make appropriate recommendations to 'responsible' persons regarding upgrading the equipment to comply with current regulations, standards and codes of practice.

Appliances must be serviced in accordance with specific manufacturer's instructions and the appropriate sections of the Gas Safety (Installation & Use) Regulations to satisfy obligations that arise as a result of the Health and Safety at Works Act.

The Gas Safety (Installation & Use) Regulations (26) demand that:

"Where a person performs work on a gas appliance (*the definition of 'work' includes maintenance*) he shall immediately thereafter examine:

a) the effectiveness of any flue

b) the supply of combustion air

c) its operating pressure or heat input or, when necessary, both

d) its operation so as to ensure its safe functioning

and forthwith take all reasonable practicable steps to notify any defect to the responsible person and, where different, the owner of the premises in which the appliance is situated, ... etc."

The aforementioned essential tasks must therefore be included in any servicing regime associated with manufacturer's specified tasks for a particular appliance.

The process of servicing may follow a similar programme to that defined in the previous commissioning section, i.e. planning, inspection, activation period, operation period and a completion period.

The activation and operation period would now include those additional tasks required for the correct servicing of the appliance.

The following checklist may provide guidance for general servicing tasks in the absence of manufacturer's information. The checklist itself is not intended to be concise or complete for any particular appliance.

Pre-Service Checks

- Check with the appliance user to establish is they are aware of any problems or faults that may already exist.

- Visually inspect the appliance and its general installation in situ to identify:
  - there is no evidence of malfunction, e.g. spillage, soot deposits, etc.
  - there is no obvious evidence of physical damage, e.g. corrosion, faulty wiring, broken components, etc.
  - the appliance is sited correctly – there is adequate clearance from combustible material and for correct maintenance (burner withdrawal), etc.
  - the appliance is adequately supported or suspended – for suspended air units the operating height is correct
  - the appliance is correctly assembled and fitted with appropriate means of isolation (gas, electricity, water)
  - there is adequate ventilation – refer to British Standard if no manufacturer's instructions are available to establish the required free area
  - the flue system is appropriate, is constructed of correct materials, is both routed and terminated correctly
  - there are no signs of water leakage from boiler sections
  - the gas connections appear to be of suitable size, appropriately labelled and of correct materials.

- With the appliance operating inspect:
  - the flame picture – does it indicate complete or incomplete combustion, does it appear to 'lift off'
  - flame ignition is correct – smooth ignition (not delayed)
  - operational controls appear to function correctly – thermostats, time switches, fan controls and safety limit devices (overheat thermostats)
  - any control link to the appliance – e.g. exhaust fan shuts off the appliance in the event of air flow failure.

Full Service

The following tasks should be carried out as required using appropriate tools, cleaning methods and agents.

- Isolate gas, electricity and where applicable, water supplies.

- Dismantle the appliance as necessary and clean dust and deposits from within the appliance casing.

- Check for signs of internal damage – to wiring or components. Clean and rectify as necessary.

- Remove main burner and injectors. Inspect and clean or replace as necessary (inspect, clean or replace ceramic plaque on radiant heaters).

- Remove pilot burner and injectors. Inspect and clean or replace as necessary.

- Where applicable, remove, clean and lubricate any exhaust or burner fan. Check air flow proving device (flow switch, pressure switch and connecting tubes).

- On warm air units, check, clean and lubricate main air fan(s) as necessary. Check any fan belt(s) for correct tension and condition.

- Inspect and clean as necessary combustion chamber, heat exchanger and flueways. Carry out flue flow test.

- Check all appliance seals and renew as necessary.

- Check operation of gas taps/valves and lubricate as necessary.

- Check all gas taps/valves/safety shut off valves for 'let-by'. Repair or replace as necessary.

- Check safety shut off valve(s)/solenoid valves/multi-functional valves for correct operation.

- Check operation of gas regulator(s) – ensure vent hole is clear and diaphragms are undamaged (no smell of gas at vent hole when in operation).

- Check all gas pipework connections and disturbed joints for tightness.

- Inspect, clean or replace as necessary any ignition device.

- Inspect, clean or replace as necessary any flame monitoring device.

- Re-assemble all components and light the appliance. Check for reliable flame ignition and monitoring (pilot and main flame).

- Check burner operating pressure is correct – in accordance with data plate and/or manufacturer's instructions.

- Check appliance input pressure is correct – investigate, report or action as necessary any over-pressure or under-pressure situation.

- Carry out flue gas analysis and/or atmosphere test – adjust as necessary for optimum and safe combustion

- Carry out spillage test.

- Check operation of controls as fitted, including:
    - appliance thermostat
    - room thermostat
    - frost thermostat
    - overheat thermostat
    - cylinder thermostat
    - safety valve
    - fan switch.

- Check system pressure on a sealed central heating system.

- Also on central heating system check ball valve in feed and expansion tank.

- Leave the appliance in working order.

- Complete a servicing report and submit to 'responsible person'. Note on the report any defects and 'Not to Current Standard' situations identified – any 'At Risk' or 'Immediately Dangerous' situation would be subject to separate action as appropriate.

GENERAL FAULT FINDING

FAULT FINDING CHECKLIST FOR COMMERCIAL APPLIANCES WITH NATURAL DRAUGHT BURNERS AND THERMO-ELECTRIC F.F.D.

Symptom	Possible Cause	Action
1. Pilot will not light	a) Gas not turned on	a) Turn on the gas at the gas service cock and/or the main gas line
	b) Air in gas line	b) Purge the gas line by depressing and holding the ON button. This may take 2-3 minutes
	c) Incorrect pilot lighting procedure	c) Follow the lighting instructions or refer to the 'Instructions for Use'
	d) Electrode lead not connected to the rear of the spark igniter	d) Reconnect the electrode lead
	e) Incorrect spark gap	e) Refer to manufacturer's instructions for correct spark gap and adjust
	f) Current tracking to earth	f) Check the electrode lead is routed clear of all metal parts
	g) Faulty spark generator or electrode assembly	g) Clear injector by blowing through or replace with new pilot injector. Clean the lint filter if fitted
	h) Pilot injector blocked	h) Replace faulty components
2. Pilot lights but will not remain alight	a) Thermocouple connections loose	a) Tighten the connections

Symptom	Possible Cause	Action
	b) Pilot flame unstable or too small to heat the thermocouple tip	b) Pilot injector partially blocked. Clear by blowing through or replace. Check the gas inlet pressure is correct – 20 mbar. Clean the lint filter if fitted
	c) Thermocouple worn out or damaged	c) Replace the thermocouple
	d) Faulty magnet unit in the flame safety device	d) Replace flame safety device
	e) Excessive draughts due to faulty seals	e) Replace seals
	f) Overheat limit thermostat tripped	f) Reset
3. Pilot established but main burner will not light	a) Main gas not turned on	a) Turn main burner valve on
	b) Gas inlet pressure low	b) Check gas filter clear. Check pressure at meter and if incorrect contact the gas supplier
	c) Controls not calling for heat	c) Check all controls are in 'on' position and work correctly
	d) No electrical supply	d) Check electricity is on. Check all fuses and replace if necessary
	e) If appliance is fitted with exhaust fan	e) Check fan is operating. Check fan switch is operating
	f) Downdraught detector or overheat limit thermostat tripped	f) Reset
	g) Faulty wiring to gas valve	g) Inspect wiring, trace circuit and replace as necessary
	h) Faulty solenoid on gas valve	h) Replace solenoid or complete valve

Symptom	Possible Cause	Action
	i) Faulty gas valve (seating stuck)	i) Replace valve
	j) Gas regulator inoperable (seating stuck)	j) Replace regulator
4. Main burner established but low fire flame only (low gas pressure)	a) Inadequate supply pressure	a) Check incoming pressure is correct (21 mbar). Report to gas emergency service if incorrect
	b) Inadequate pipe size to appliance	b) If supply pressure is correct check pipe diameter is correct for required flow. Replace undersized pipe with correct size
	c) Gas pipe to appliance partially blocked	c) Investigate pressure at points along gas system. Renew any blocked section
	d) Gas regulator not working correctly	d) Check vent hole is not blocked and clean as necessary
	e) Gas valve/regulator filter partially blocked (multi-functional valve)	e) Strip out valve/regulator and clean inlet
	f) If two stage gas valve fitted, faulty valve (2nd stage)	f) Replace valve
5. Main burner established but then cuts out	a) Heater/boiler overheat limit activated Warm air fan not working Air fan belt(s) slipping or broken Boiler circulating pump not operating	a) Reset Replace motor and/or fan Tighten or replace fan belt(s) Remove pump and repair/replace as necessary
	b) Pilot gas starvation, poor thermocouple signal	b) Check operating pressures (see 2b, 2c and 4a-d)

Symptom	Possible Cause	Action
6. Burner flame, lifting, sooting, smothering, smelly	a) Insufficient air for combustion	a) Check for obstruction at combustion air intake. Check for ventilation air supply to compartment or room. Check grille sizes, etc.
	b) Wrong burner or jets fitted. Incorrect gas in use for burner design	b) Disconnect gas supply. Check manufacturer's literature. Fit correct burner/jets for gas supply (i.e. natural gas, LPG)
	c) Sooting/smell:	c)
	caused by blocked or partially blocked flue	Carry out flue test, remove obstruction and retest
	broken jets or faulty burner	Observe burner operating before stripping down; faulty jets are not always revealed except under working conditions. Replace faulty jets/burner
	burner out of position	Check burner location brackets for damage, corrosion, etc. Clean heat exchanger, primary flue. Relocate burner
	recirculation of products	Check on installation for evidence of restriction on ventilation that would cause burner to pull on draught diverter. Check flue system, carry out flue test, CO/CO_2 test. Check ventilation and air supply
	linting	Remove lint from burners/screens
	oversized injectors	Clean combustion chamber, replace injector(s) with correct size, reset to correct gas rate

Symptom	Possible Cause	Action
	gas rate set too high	Clean combustion chamber, reset to correct gas rate, check operation
	faulty appliance governor failing to maintain correct gas pressure	Strip governor, check governor spring (where fitted), clean valve and seating, clean diaphragm or replace as necessary. Note: oil leather diaphragms only. Reassemble, refit and check for correct working. Adjust as necessary
	products not clearing due to badly positioned flue; broken or no terminal fitted	Check flue, renew or re-route to specification
	sooting; air leakage upsetting balance conditions (room-sealed appliances)	Look for reason for air leakage – faulty gasket, etc.

FAULT FINDING CHECKLIST FOR COMMERCIAL APPLIANCES WITH FORCED/INDUCED DRAUGHT BURNERS AND AUTOMATIC CONTROL

Symptom	Possible Cause	Action
7. Burner will not start (control box indicator in shut down mode)	a) No electrical supply	a) Check electricity is switched on. Check all fuses and replace as necessary
	b) Controls not calling for heat (timeswitch, thermostat, etc.)	b) Check all external controls are in 'on' position and work correctly
	c) Low gas pressure (CE marked burners)	c) Ensure gas is turned on. Check supply pressure is adequate for LP gas switch
	d) If appliance is fitted with exhaust fan	d) Check fan is operating. Check fan switch is operating

Symptom	Possible Cause	Action
	e) Overheat limit thermostat tripped	e) Reset
	f) Faulty wiring to burner	f) Inspect wiring. Trace circuit. Renew as necessary
8. Burner will not start (control box indicator in 'lock-out' mode)	a) Variable	a) Note the position on control box indicator. Reset. Observe burner lighting sequence
9. Burner will not start (control box indicator goes immediately to 'lock-out')	a) Flame simulation, faulty UV detector (where fitted)	a) Isolate electricity and replace UV detector
10. Burner will not start (control box indicator rotates continually but does not activate burner)	a) 'No air' position on air pressure switch not proved	a) Check fan is not running (possibly due to flue draught). Check air pressure switch is in 'open' position (no air)
	b) Air pressure switch contacts welded in closed position	b) Replace pressure switch
	c) Electrical wiring fault	c) Inspect wiring. Trace circuit. Renew as necessary
	d) Faulty control unit	d) Replace control unit
11. Burner will not start (control box indicator stuck on purge – no electricity)	a) Seized burner fan motor	a) Replace fan motor with new. Check fuse in control box (where fitted) and replace as necessary
	b) Burner fan motor – electrical short	b) Rewire or replace motor with new. Check fuse in control box (where fitted) and replace as necessary
	c) Motor overload tripped (three phase electrical supply)	c) Reset

Symptom	Possible Cause	Action
12. Burner fan runs for short time and then stops (control box indicator goes to 'lock-out'	a) Low air pressure b) Air pressure switch contacts welded in closed position c) Flame simulation, faulty UV detector (where fitted) d) Open circuit on gas safety shut off valve	a) Check air pressure switch setting is correct – adjust as necessary. Check burner fan is running correctly b) Replace pressure switch c) Isolate electricity and replace UV detector d) Replace valve
13. Burner fan continues to run for longer than required purge time (control box indicator stopped on purge position). **High/low or modulating burner only**	a) High fire position air damper micro-switch not made b) Low fire position air damper micro-switch not made	a) Check and adjust micro-switch as necessary b) Check and adjust micro-switch as necessary
14. Burner fan runs for full purge cycle. Start gas does not ignite (control box indicator goes to 'lock-out'	a) Gas not turned on b) Air in gas line c) Insufficient gas pressure d) Too much air for combustion e) No spark – faulty electrode f) No spark – incorrect spark gap	a) Turn on the gas at the service cock and/or the appliance b) Purge the gas line c) Check start gas pressure and adjust as necessary d) Adjust air damper as necessary e) Inspect electrode for evidence of damaged ceramic or tracking of spark. Replace electrode f) Check gap, adjust as necessary

Symptom	Possible Cause	Action
	e) Overheat limit thermostat tripped	e) Reset
	f) Faulty wiring to burner	f) Inspect wiring. Trace circuit. Renew as necessary
8. Burner will not start (control box indicator in 'lock-out' mode)	a) Variable	a) Note the position on control box indicator. Reset. Observe burner lighting sequence
9. Burner will not start (control box indicator goes immediately to 'lock-out')	a) Flame simulation, faulty UV detector (where fitted)	a) Isolate electricity and replace UV detector
10. Burner will not start (control box indicator rotates continually but does not activate burner)	a) 'No air' position on air pressure switch not proved	a) Check fan is not running (possibly due to flue draught). Check air pressure switch is in 'open' position (no air)
	b) Air pressure switch contacts welded in closed position	b) Replace pressure switch
	c) Electrical wiring fault	c) Inspect wiring. Trace circuit. Renew as necessary
	d) Faulty control unit	d) Replace control unit
11. Burner will not start (control box indicator stuck on purge – no electricity)	a) Seized burner fan motor	a) Replace fan motor with new. Check fuse in control box (where fitted) and replace as necessary
	b) Burner fan motor – electrical short	b) Rewire or replace motor with new. Check fuse in control box (where fitted) and replace as necessary
	c) Motor overload tripped (three phase electrical supply)	c) Reset

Symptom	Possible Cause	Action
12. Burner fan runs for short time and then stops (control box indicator goes to 'lock-out'	a) Low air pressure	a) Check air pressure switch setting is correct – adjust as necessary. Check burner fan is running correctly
	b) Air pressure switch contacts welded in closed position	b) Replace pressure switch
	c) Flame simulation, faulty UV detector (where fitted)	c) Isolate electricity and replace UV detector
	d) Open circuit on gas safety shut off valve	d) Replace valve
13. Burner fan continues to run for longer than required purge time (control box indicator stopped on purge position). **High/low or modulating burner only**	a) High fire position air damper micro-switch not made	a) Check and adjust micro-switch as necessary
	b) Low fire position air damper micro-switch not made	b) Check and adjust micro-switch as necessary
14. Burner fan runs for full purge cycle. Start gas does not ignite (control box indicator goes to 'lock-out'	a) Gas not turned on	a) Turn on the gas at the service cock and/or the appliance
	b) Air in gas line	b) Purge the gas line
	c) Insufficient gas pressure	c) Check start gas pressure and adjust as necessary
	d) Too much air for combustion	d) Adjust air damper as necessary
	e) No spark – faulty electrode	e) Inspect electrode for evidence of damaged ceramic or tracking of spark. Replace electrode
	f) No spark – incorrect spark gap	f) Check gap, adjust as necessary

Symptom	Possible Cause	Action
	g) No spark – faulty wiring	g) Inspect wiring for evidence of damage or tracking of spark. Replace as necessary
	h) No spark – faulty spark generator	h) Check electrical supply from control box and/or generator output spark. Try replacement generator
	i) Start gas safety shut off valve fails to open	i) Check wiring and establish power to valve. Replace solenoid or complete valve as necessary
15. Burner lights on start gas but goes out after three or five seconds (control box indicator goes to 'lock-out')	a) Incorrect gas/air mix for stable combustion	a) Check and adjust air damper and/or gas pressure
	b) No flame signal – damaged flame electrode (flame ionisation)	b) Check electrode is correctly site. Inspect electrode for evidence of damaged ceramic. Replace electrode
	c) No flame signal – damaged/dirty/faulty UV detector	c) Check UV detector is correctly sited. Clean UV cell. If fault persists, replace UV detector
	d) No flame signal – damaged or faulty wiring	d) Inspect wiring for evidence of damage. Replace as necessary
	e) Faulty amplifier in control box	e) Replace control box
16. Burner start gas flame established, but does not go to main gas stage (control box indicator may go to 'lock-out' or proceed to 'run' stage, dependent on manufacturer and configuration)	a) No main gas	a) Turn on main gas cock
	b) Incorrect main gas/air mix for stable combustion	b) Check and adjust air damper and/or gas pressure
	c) Main gas safety shut off valve fails to open	c) Check wiring and establish power to valve. Replace solenoid or complete valve as necessary

Symptom	Possible Cause	Action
	d) External temperature controller not working	d) Check controller setting is correct and working correctly
17. Gas flame established, burner in 'run' mode, poor combustion analysis results, unstable flame	a) Incorrect air/gas ratio b) Damaged burner head/ diffuser plate	a) Adjust as necessary b) Remove burner and inspect/repair/replace

FAULT FINDING CHECKLIST FOR COMMERCIAL AIR HEATERS

Symptom	Possible Cause	Action
18. Fan runs continuously after main gas is off	a) Faulty fan control b) Fan control settings incorrect c) Auto/manual switch set to manual d) Faulty auto/manual switch	a) Fan control stuck on – replace b) Check fan control 'on' setting, alter to lower temperature to check switch. Try raising 'off' setting, reduce differential. Replace if required faulty fan control c) Switch to auto – if no response, disconnect switch from circuit, test as for faulty auto/manual switch below d) Use ohms x 1 range of meter to check switch operation and replace if necessary. Leave switch connecting wires disconnected and taped up separately, so that heater may be used on auto only if replacement switch is not at hand. Advise customer

Symptom	Possible Cause	Action
19. Fan runs but main gas cuts in and out frequently (cycling)	a) All warm air outlets shut causing high unit temperature and overheat thermostat operation	a) Advise customer on use of warm air outlets
	b) Fan speed too slow causing high unit temperature and overheat thermostat operation	b) Report to customer, do not attempt to adjust fan speed. Random adjustment may cause other problems
	c) Overheat thermostat set too low, too close to, or overlapping 'fan on' control setting	c) Adjust as necessary
	d) Return air path obstructed	d) Clear obstruction, advise customer if fault due to customer action
	e) Blocked filter	e) Advise customer to clean if cleanable pattern or to replace if expendable type. Operate heater temporarily without filter
	f) Room thermostat anticipator heating setting (series heater)	f) First try to raise setting, keeping within the manufacturer's recommended range of adjustment. If this does not effect a cure, raise to a slightly higher setting
	g) Room thermostat anticipator faulty	g) See flowchart 17 – switches and thermostats
	h) Dirty or clogged fan blades	h) Clean as necessary

Symptom	Possible Cause	Action
20. Main burner shuts down in response to room thermostat being satisfied but fan maintains a continuous cold air flow	a) Fan control 'off' setting too low. Where the fan control 'off' setting is close to or below that of the air returned to the heater, there will be long periods when delivery temperatures, further reduced by ductwork losses, will be much below acceptable comfort standards b) Faulty fan control	a) Try raising 'off' setting 10°F (5.5°C), reduce differential slightly if necessary b) Replace
21. When time switch goes to 'off' sequence, both fan and burner immediately switch off	a) Faulty wiring	a) Check with wiring diagram and correct as necessary. Check fan delay arrangements and time switch wiring. Rewire as necessary
22. Main burner on, but fan cycles on and off frequently on call for heat	a) Excessive fan speed. Burner lights in response to call for heat, fan 'on' temperature is reached; blow starts but being excessive soon causes temperature drop in heat exchanger until fan 'off' operates. This cycle then repeats again and again b) Low gas rate. A low gas rate will cause this effect by preventing sufficient temperature rise to be obtained across heater	a) Fan speed adjustment is necessary to increase temperature rise across heater to recommended figure - 80°F to 90°F (44°C to 50°C). Report; do not carry out adjustment, other design factors may be involved Try raising 'off' setting 10°F (3-5°C). Reduce differential slightly if necessary b) Check and cure reason for low gas rate
23. Limit control cuts out before fan control starts fan	a) Fan differential is set too high. Temperature reached in heat exchanger before blow can start is so high that overheat control operates	a) Reduce differential setting of fan control

Symptom	Possible Cause	Action
	b) Faulty overheat control	b) Faulty calibration or faulty component – replace
	c) Faulty fan control	c) Replace
24. Fan works on and off intermittently after main burner shuts down in response to thermostat switch or clock	a) Fan control out of adjustment. On certain installations, particularly where the fan control 'off' temperature is set too high, and with a small differential setting, cool air blown down through the heater by the fan immediately following main burner shutdown may prematurely cool the small mass of the control sensing element to cause early fan 'off' operation. Residual heat remaining in the heater reheats the fan control element to a temperature high enough to cause another fan 'on' operation. This 'on'/'off' cycle may occur several times on each occasion that the main burner shuts down. The process then repeats	a) Try to increase the differential or to lower the fan 'off' temperature setting. (Remember that too low a fan 'off' setting may promote a complain of cold blow)
25. Burner flames not steady – lifting, sooting, smothering	a) Air leaks or perforations between combustion and fan ducts (warm air heaters)	a) View flame picture with fan off. Repeat and view with fan operating. Now repeat and view with fan on with all warm air outlets closed temporarily. If flame picture is worsened by the last two tests this is a sure indication of leakage. Disconnect appliance. Repair or renew

Symptom	Possible Cause	Action
	b) Low gas pressure:	b)
	poor supplies	Check standing and working pressure. If poor supply or service, clear
	supplies undersized	If supplies obviously undersized or excessively long, advise customer. Renew
	governor faulty or pressure setting incorrect	Check at appliance governor, clean, adjust, repair or replace as necessary. If a range-rated heater, up-rate as necessary. Do not exceed maximum rate for heater
	obstruction at burner ports, etc.	Check for obstruction at burner ports, injectors, etc. Clean as required, re-check gas rate
26. Insufficient heat in all rooms at all times	a) Faulty room thermostat	a) Intermittent operation – dirty contacts, clean or replace. Grossly out of calibration, adjust or replace. Wired incorrectly
	b) Room thermostat not level (mercury type)	b) Affects calibration. Level and adjust as necessary
	c) Room thermostat badly positioned	c) Advise customer on re-positioning and re-site thermostat
	d) Insufficient return air:	d)
	from rooms	Clear any obstruction. Advise customer
	to heater; may be due to obstruction or lack of provision in design	Remove obstruction. Advise customer of design fault

Symptom	Possible Cause	Action
	e) Whole premises over-ventilated	e) Find cause and advise customer accordingly. If due to opening of windows, advise that this is best done during clock-controlled shut-down periods and not when heater is in operation
	f) Blocked air filter (this may be indicated by intermittent operation of the limit stat)	f) Clear obstruction
	g) Undersized unit	g) If this is judged to be the root cause of the complain, report to customer
	h) Ducts not lagged	h) Advise customer
	i) Fan speed too slow	i) Check condition of air filter (where fitted). Fan speed adjustment is required to give recommended temperature rise through heater of between 80°F to 90°F (44°C to 50°C) checked at return air grille and nearest warm air outlet to heater, with all warm air outlets in normal operational position. This should mean that with a temperature at the return air grille of 65°F the outlet temperature at the nearest warm air outlet will be between 145°F to 155°F (63°C to 69°C). It is not recommended that random adjustments be made of fan speeds – other factors may be affected by this action

Symptom	Possible Cause	Action
		(With a slow fan speed the actual temperature at the warm air outlet may be quite high – in fact higher than normal. The complaint arises because the warmed air is not being projected to where it is wanted. In addition, the situation may be aggravated by the intermittent operation of the overheat stat due to high temperatures reached in the heat exchanger)
		Do not attempt to regulate fan speed – other factors may be affected
		The desired ambient room temperature may be reached, but the high velocity of the warmed air causes discomfort and a sense of feeling cold.
27. Insufficient heat with excessive 'on'/'off' cycling of heater	a) Overheat thermostat operating – blocked filter	a) Check for blocked filter (where fitted). Check operation of heater without filter; if now working satisfactorily (and cleanable filter is fitted), advise customer to clean. If expendable type of filter is fitted advise customer to replace. The heater may be run without a filter until a replacement is available

Symptom	Possible Cause	Action
	b) Warm air outlets closed	b) Check if all warm air outlets and/or dampers are in closed position – advise customer always to leave one or more open
	c) Fan motor faulty	c) Check out fan motor, replace if faulty
	d) Fan motor set to run too slow	d) Report; do not attempt to adjust fan speed or other factors may be affected
	e) Broken or slack belt (where fitted)	e) Adjust belt to leave 2 cm-2.5 cm of downward free movement when tested by hand midway between pulleys. Replace broken belt
	f) Insufficient return air arriving at heater	f) (Non-connected return air duct systems only.) Check for obstructions between compartment return air grille and heater intake Check for observations of return air arrangements from heated rooms
	g) Gas rate	g) Fan speed may be correct and all else satisfactorily designed, but if gas rate is too high, fast heat build up in heat exchanger may cause 'overheat' stat to operate. May be the result of uprating range-rated heaters without taking account of fan speed setting

Note: All indicators above cause overheat shutdown. All are events that tend to slow down air flow through heater, causing high temperatures within the heat exchanger and overheat stat operation. Meanwhile, the room stat is calling for heat; hence the continuous cycling 'on'/'off' effect.

Symptom	Possible Cause	Action
	h) Faulty overheat stat	h) **Automatic reset.** Try to run heater from cold with overheat stat contacts temporarily bridged across. If heater runs satisfactorily, replace stat; if not, search for fault elsewhere. Test out heater **Manual reset.** A manual reset button may be fitted to certain overheat stats; try resetting. If stat operates to 'off' again, try bridging. If heater now runs satisfactorily, replace stat; if not, search for fault elsewhere
	i) Faulty fan control	i) Raise 'off' setting 10°F (5.5°C), reduce differential slightly as necessary. Replace if faulty
28. Noisy operation – buzzing or chattering sound	a) Due to solenoid armature chatter	a) Check location – armature of most solenoids must be vertical. See manufacturer's literature
	b) Faulty solenoid coil	b) Disconnect from circuit, check d.c. resistance if known. If outside tolerance of 10% of normal value – replace coil
	c) Operating voltage low	c) Where transformer operated at low voltage, check that voltage tapping on primary is correct. Check for faulty wiring, frayed insulation, dirty terminals, etc. Make good or replace as necessary

Symptom	Possible Cause	Action
	d) Dirt or other foreign matter on armature	d) If control can be dismounted (e.g. Sperryn or similar type), dismantle, clean and adjust, reassemble and test out. If Sperryn 'triangular' shaped valve head with plain armature, exchange for 'centre popped' armature. If not available a Sperryn SS.67 diaphragm protected valve or recommended alternative should be fitted
	e) Faulty rectifying diodes in solenoid	e) Replace solenoid
	f) Loose armature cover	f) Tighten securing screw
29. Noisy motor fan	a) Motor or fan mounting faulty	a) Tighten or replace anti-vibration mounting as necessary
	b) Fan belt worn, too slack or too tight	b) Adjust or replace as necessary
	c) Motor and fan pulleys out of alignment	c) Adjust by repositioning pulleys on shafts. Adjust belt if necessary
	d) Too much end play on motor shaft:	d)
	with no provision for adjustment	Check if motor end cover retaining bolts loose; tighten and re-check for end play
	with provision for adjustment	Where end play adjustment is provided, i.e. by means of brass collars as on certain Halcyon heaters, adjust collars as necessary. (A 3/16" Allen key is required)

Symptom	Possible Cause	Action
	e) Bent fan shaft or motor shaft	e) Replace fan or motor as appropriate – there is no other remedy
	f) Motor fan blades bent, fouled or out of balance	f) Fan blades may be carefully realigned in some cases. Build up of dust or sediment on blades should be carefully cleaned off
30. Explosive ignition or noisy extinction of main burner	a) Explosive ignition or noisy extinction of main burner	a) Check working pressure, check pilot filter. Replace if necessary, check pilot burner for displacement, relocate and secure

FAULT FINDING CHECKLIST FOR COMMERCIAL BOILERS

Symptom	Possible Cause	Action
31. Noisy operation of system	a) Solenoid valve chatter	a) Check location – armature of most solenoids must be vertical
	b) Noisy flame ignition	b) Check working pressure, check pilot filter, clean or replace
	c) Noisy flame extinction	c) Check working pressure, check burner, clean or replace
	d) Circuit noise: pump pressure appears excessive	d) Take note of present setting, adjust to lower setting; if noise lessened but system performing unsatisfactorily, report to supervisor

Symptom	Possible Cause	Action
	incorrect pump position in circuit, pumping over, air drawn into circuit	See section on pump positions; check position in circuit in relation to vent and feed pipes and that pressure excessed by feed and expansion tank (state pressure). Replace or redesign
	e) Cavitation noises in pump	e) Vent pump; look for cause in relation to pump position
	f) Kettling or bumping in appliance (low water content boilers)	f) Bypass loop closed or absent, or simply residual heat after shutdown causing local boiling; sometimes accompanied by nuisance tripping of overheat stat. Fit bypass loop
	g) Crackling noises on heating up or cooling down	g) Due to lack of clearance for installed pipework which expands and contracts with temperature exchanges. No simple cure. May sometimes occur on conventional high-water content cast iron boilers, where due to the positioning of the flow and return off-take tappings, local boiling occurs in 'pockets' Sometimes a cure can be effected by repositioning the boilerstat phial
32. Burner flame not steady	a) Air leaks in combustion chamber on balanced flue models	a) Check seals – renew as necessary

Symptom	Possible Cause	Action
	b) Gas rate set too high – smothering	b) Adjust gas rate – do not exceed maximum rating on range-rated appliances
	c) Flue pull affected by room ventilation fan. (Conventionally flued appliance)	c) Advise customer – additional ventilation required
	d) Balanced flue inlet and outlet ducts not correctly fitted or moved out of position. Flue terminal missing	d) Realign and reseal or arrange for replacement
	e) Appliances sited in line of strong draught. (Conventionally flued appliance)	e) Inform customer, fit draught diverter shield, try change of terminal, re-route flue
	f) Fanned flue appliance	f) Check for correct operation of fan
33. Boiler overheating	a) Solenoid valve sticking in 'on' position	a) Dismantle valve, clean valve and seat, reassemble and test
	b) Faulty boilerstat	b) Check boiler and stat for operation, calibration and renewal if faulty
	c) Boilerstat phial out of position	c) Check that boilerstat phial is correctly positioned in phial pocket on boiler

BES PUBLICATIONS

Building Engineering Services continue to provide the gas, electric, water and refrigerant industries with a range of popular, respected and competitively priced publications.

These publications can be used either as the basis of training or for reference in the workplace. Some can also be used for assessment purposes. All are published in A4 format, with the most popular also available as A5, pocket-sized books.

DOMESTIC GAS

GAS SAFETY (G1)
Format: A4 in a ringbinder

The complete manual for reference or self-study. All of the essentials in 300 pages, with clear explanations and illustrations, covering ◆gas pipework ◆gas supply ◆combustion ◆appliance gas safety devices and gas controls ◆principles of gas flues ◆flueing standards ◆ventilation requirements ◆emergency procedures ◆unsafe situations ◆warning notices and labels. Also included is the HSE publication ◆*Safety in the installation and use of gas systems and appliances* (G31) which covers the HSE Gas Safety (Installation and Use) Regulations 1998 – Approved Code of Practice and Guidance, a ◆*Course Workbook* and a booklet of ◆*Practical Tasks* for you to complete.

GAS SAFETY (G2)
Format: A5 Wiro-bound

All the information and diagrams from *GAS SAFETY (G1)* in a handy size for reference on the job and for carrying in the service van.

DOMESTIC GAS APPLIANCES (G5)
Format: A4 in a ringbinder

Contains all seven of the domestic natural gas appliance manuals from ConstructionSkills in one package, plus the *Domestic Natural Gas Appliances Course Workbook (G14)*. The easy-to-use format makes it ideal for engineers working with a range of domestic appliances.
Each manual can also be purchased individually:

- Heating Boilers/Water Heaters (G7)
- Cookers (G8)
- Ducted Air Heaters (G9)
- Fires and Wall Heaters (G10)
- Tumble Dryers (G11)
- Meters (G12)
- Instantaneous Water Heaters (G13)

DOMESTIC GAS APPLIANCES (G6)
Format: A5 Wiro-bound

All the information and diagrams from the *DOMESTIC GAS APPLIANCES (G5)* in a handy size for reference on the job and for carrying in the service van.

FAULT-FINDING TECHNIQUES (G17)
Format: A4

Problems with locating that elusive fault? Follow the step-by-step techniques in this hands-on manual and speed up your fault finding on central heating systems.

SAFETY IN THE INSTALLATION AND USE OF GAS SYSTEMS AND APPLIANCES (G31)
Format: A4

An essential HSE publication for all those working with domestic gas. It gives advice on how to comply with *The Gas Safety (Installation and Use) Regulations 1998 – Approved Code of Practice and Guidance,* which has a special legal status. For example, if you are prosecuted for breach of health and safety law, and it is proved that you have not followed the relevant parts of the Code, a court will find you at fault (unless you can show that you have complied with the law in some other way).

COMMERCIAL AND INDUSTRIAL GAS

COMMERCIAL GAS SAFETY (G88)
Format: A4 in a ringbinder

An essential training and reference manual for those working in the commercial environment. It includes key sections from the popular GAS SAFETY (G1) and incorporates information from two other commercial publications (G23 and G24) which can be purchased separately) making this the definitive training and reference manual for commercial work. It covers ◆commercial gas safety ◆pipework and ancillary equipment ◆gas pipework ◆gas supply ◆combustion ◆appliance gas safety devices and gas controls ◆principles of gas flues ◆flueing standards ◆ventilation requirements ◆emergency procedures ◆unsafe situations ◆warning notices and labels. Also included is the HSE publication ◆*Safety in the installation and use of gas systems and appliances* (G31) and ◆*Course Workbooks* and *Practical Tasks* (G3, G4, G83 and G84).

COMMERCIAL GAS SAFETY (G23)
Format: A4

An essential supplement for engineers working in the commercial environment. If you already own a *GAS SAFETY (G1)* pack, all you need is this book with its commercial gas-specific sections ◆combustion and flue gas analysis ◆burners ◆controls and control systems ◆flues ◆ventilation ◆pressure and flow.

COMMERCIAL PIPEWORK AND ANCILLARY EQUIPMENT (G24)
Format: A4

An essential guide for engineers working on commercial pipework, with clear information on ◆pipework design ◆soundness testing and purging ◆commercial metering ◆boosters and compressors.

COMMERCIAL APPLIANCES (G25)
Format: A4

A comprehensive guide to the installation and commissioning of direct and indirect fired appliances, radiant heating and gas equipment.

COMMERCIAL CATERING (G26)
Format: A4

Essential information on installing, commissioning and servicing commercial catering appliances.

To obtain further information and order any of the publications listed, contact Publications on: Tel: 01485 577800 / Fax: 01485 577758 / E-mail: publications@cskills.org / www.cskills.org/publications

LIQUEFIED PETROLEUM GAS (LPG)

LIQUEFIED PETROLEUM GAS SAFETY (G80) Format: A4 in a ringbinder/A4
The industry reference manual for those working only on LPG systems. It covers all you need to know about ✦combustion ✦appliance gas safety devices and gas controls ✦principles of gas flues ✦flueing standards ✦ventilation requirements ✦emergency procedures ✦unsafe situations ✦warning notices and labels.
This pack consists of: ✦*Gas Safety (G1) pack*, ✦*Liquefied Petroleum Gas Safety (G18) book*, ✦*Liquefied Petroleum Gas Safety Course Workbook (G81)*.

LIQUEFIED PETROLEUM GAS SAFETY (G18) Format: A4
The essential bolt-on to those working with natural gas and looking to extend into LPG. If you already own a *GAS SAFETY (G1)* pack, all you need is this book with its LPG-specific sections ✦installation ✦fire precautions and procedures ✦combustion ✦testing and commissioning installations ✦service pipework ✦bulk gas supply systems ✦the leisure industry.

ELECTRICAL

BS 7671: REQUIREMENTS FOR ELECTRICAL INSTALLATION (E1) Format: A4 Wiro-bound
The standard reference book for electrical work. The easy-to-follow text, supported by diagrams, explains the complex regulations in terms a practical electrician can understand. It now incorporates reference to the IEE on-site guide that enables you to make calculations and design circuits in a much quicker and simpler manner.

ELECTRICAL INSTALLATION PACK (E3) Format: A4 in a ringbinder
Over 430 pages of illustrated reference material divided into four sections:

- Basic Practical Skills – describes the tools required for electrical installation work and how to use them
- Wiring Installation Practice – deals with terminating cables, flexible cords and installing PVC cables, conduit trunking, MICC, SWA and FP200 wiring systems. (Complies with the 16th Edition *IEE Wiring Regulations*)
- Basic Electrical Circuits – covers standard circuit arrangements for lighting and power circuits, and relevant IEE Regulations
- Safety at Work – essential advice on safety at work, from securing ladders to dealing with electric shock. It also gives the key points of relevant Acts and Regulations.

ESSENTIAL ELECTRICS (E14) Format: A4
An indispensable reference book for plumbers, gas fitters and heating and ventilating engineers whose work requires basic electrical knowledge and an understanding of electrical regulations.

CENTRAL HEATING CONTROLS (E15) Format: A4
Deals with different types of central heating control systems for wiring and fault finding.

COMBINATION BOILERS (E19) Format: A4
An invaluable reference manual for engineers who want to understand the principles of combination boilers. This manual covers most of the content for the ConstructionSkills Essential Electrics and Combination Boiler Fault Finding course. Over 80 pages of illustrated reference information covering ✦types of boilers ✦designs ✦wiring diagrams ✦installation ✦commissioning and servicing ✦fault finding.

WATER

UNVENTED HOT WATER STORAGE SYSTEMS (W2) Format: A4
An informative guide for installing unvented hot water storage systems. It covers most of the content for the ConstructionSkills training and assessment scheme, including: ✦types of system ✦design ✦controls ✦installation ✦commissioning and decommissioning ✦servicing and fault diagnosis ✦relevant Building Regulations ✦good practice.

REFRIGERANTS

SAFE HANDLING OF REFRIGERANTS (R2) Format: A4
Essential information, primarily designed for operatives undertaking ConstructionSkills Safe Handling of Refrigerants training and assessments, it covers ✦environmental impact ✦fluorocarbon control and alternatives ✦regulations ✦recovery and handling ✦refrigeration theory ✦good practice ✦automotive installations.

SAFE HANDLING OF ANHYDROUS AMMONIA (R4) Format: A4
Essential information for handling anhydrous ammonia. Primarily designed for operatives undertaking ConstructionSkills Safe Handling of Anhydrous Ammonia training and assessments, it covers ✦safety and environmental issues ✦regulations ✦good practice.

PIPEWORK AND BRAZING (R6) Format: A4
Primarily for operatives undertaking ConstructionSkills Pipework and Brazing training and assessments for refrigeration systems, it covers ✦health and safety ✦materials and equipment ✦lighting procedures.

To obtain further information and order any of the publications listed, contact Publications on: Tel: 01485 577800 / Fax: 01485 577758 / E-mail: publications@cskills.org / www.cskills.org/publications